Lin Wang
Juxiang Huang
Minghu Jiang

TERF1/USP14/CCRK/BEST1-repressive microtube in human left hemisphere

Lin Wang
Juxiang Huang
Minghu Jiang

TERF1/USP14/CCRK/BEST1-repressive microtube in human left hemisphere

LAP LAMBERT Academic Publishing

Impressum / Imprint
Bibliografische Information der Deutschen Nationalbibliothek: Die Deutsche Nationalbibliothek verzeichnet diese Publikation in der Deutschen Nationalbibliografie; detaillierte bibliografische Daten sind im Internet über http://dnb.d-nb.de abrufbar.

Bibliographic information published by the Deutsche Nationalbibliothek: The Deutsche Nationalbibliothek lists this publication in the Deutsche Nationalbibliografie; detailed bibliographic data are available in the Internet at http://dnb.d-nb.de.

Coverbild / Cover image: www.ingimage.com

Verlag / Publisher:
LAP LAMBERT Academic Publishing
ist ein Imprint der / is a trademark of
OmniScriptum GmbH & Co. KG
Heinrich-Böcking-Str. 6-8, 66121 Saarbrücken, Deutschland / Germany
Email: info@lap-publishing.com

Herstellung: siehe letzte Seite /
Printed at: see last page
ISBN: 978-3-659-67393-1

TERF1/USP14/CCRK/BEST1-repressive microtube in human left hemisphere

Lin Wang[1], Juxiang Huang[1], Minghu Jiang[2]

1 Biomedical Center, School of Electronic Engineering, Beijing University of Posts and Telecommunications, Beijing, 100876, China

2 Lab of Computational Linguistics, School of Humanities and Social Sciences, Tsinghua University, Beijing, 100084, China

@Corresponding author: Lin Wang (Prof. Dr.), Biomedical Center, School of Electronics Engineering,

Beijing University of Posts and Telecommunications, Beijing, 100876, China.

Email: wanglin98@tsinghua.org.cn Tel: 8610-13240981826

We have no conflict of interest

Table of Contents

Chapter 1: High *TERF1* inside-outside-in inhibitive estrogen-induced morphogenesis through *LGALS3BP-SELENBP1-CA2-GSTM3*

Abstract

27 different Pearson mutual-positive-correlation *TERF1*-repressive molecular network was constructed from 59 overlapping of 204 GRNInfer and 118 Pearson under *TERF1* CC $\leqslant$-0.25 in high human left hemisphere compared with low chimpanzee left hemisphere. Our identified *TERF1* inside-outside-in inhibitive molecular network showed *LGALS3BP* (lectin galactoside-binding soluble 3-binding protein), *SELENBP1* (selenium-binding protein 1), *CA2* (carbonic anhydrase II), *GSTM3* (glutathione S-transferase mu 3 (brain)) in high human left hemisphere. *TERF1* inside-outside-in inhibitive terms network includes extracellular region, extracellular matrix (sensu Metazoa), extracellular space, nucleus, membrane, cytoplasm; cellular defense response, response to estrogen stimulus; cell adhesion, morphogenesis of an epithelium; scavenger receptor activity, protein-binding, selenium-binding, carbonate dehydratase activity, zinc ion-binding, lyase activity, metal ion-binding, glutathione transferase activity, transferase activity based on integrative GO, KEGG, GenMAPP, BioCarta and disease databases in high human left hemisphere. Therefore, we propose high *TERF1* inside-outside-in inhibitive estrogen-induced morphogenesis through *LGALS3BP-SELENBP1-CA2-GSTM3* in human left hemisphere.

Keywords: *TERF1* inhibitive network; inside-outside-in; morphogenesis; estrogen

Introduction

TERF1 is one of our identified significant molecules (fold change 2) in high human compared with the corresponding low chimpanzee left hemisphere. *TERF1* appears chromosome, telomeric region, nuclear telomere cap complex, nucleus, nucleoplasm, chromosome, spindle; double-stranded telomeric DNA-binding, DNA bending activity, protein homodimerization activity, telomerase inhibitor activity; cell cycle, mitosis, negative regulation of telomere maintenance via telomerase, negative regulation of telomere maintenance via semi-conservative replication, regulation of transcription, cell division, negative regulation of telomerase activity, telomere maintenance via telomere shortening; cell cycle, M phase, M phase of mitotic cell cycle; telomeres, telomerase, cellular aging, and immortality based on GO, KEGG, GenMAPP, BioCarta and disease databases. Telomeric negtive relationship migration has been reported in references as follows: Retraction. Bone morphogenetic protein-7 induces telomerase inhibition, telomere shortening, breast cancer cell senescence, and death via Smad3; Bone morphogenetic protein-7 inhibits telomerase activity, telomere maintenance, and cervical tumor growth; Bone morphogenetic protein-7 induces telomerase inhibition, telomere shortening, breast cancer cell senescence, and death via Smad3 [1-3].

Estrogen effect on morphogenesis has been reported in references as follows: p130Cas over-expression impairs mammary branching morphogenesis in response to estrogen and EGF; Estrogen receptor beta is essential for sprouting of nociceptive primary afferents and for morphogenesis and maintenance of the dorsal horn interneurons; Estrogen receptor-alpha expression in the mammary epithelium is required for ductal and alveolar morphogenesis in mice; Paracrine signaling through the epithelial estrogen receptor alpha is required for proliferation and morphogenesis in the mammary gland; Repressor of estrogen receptor activity (REA) is essential for mammary gland morphogenesis and functional activities: studies in conditional knockout mice; Identification of an estrogen-regulated circadian mechanism necessary for breast acinar morphogenesis; Interaction with endothelial cells is a prerequisite for branching ductal-alveolar morphogenesis and hyperplasia of preneoplastic human breast epithelial cells: regulation by estrogen; Bisphenol A promotes dendritic morphogenesis of hippocampal neurons through estrogen receptor-mediated ERK1/2 signal pathway [4-11]. Yet high *TERF1* inside-outside-in inhibitive estrogen-induced morphogenesis through *LGALS3BP-SELENBP1-CA2-GSTM3* in human left hemisphere is not clear.

27 different Pearson mutual-positive-correlation *TERF1*-repressive molecular network was constructed from 59 overlapping of 204 GRNInfer and 118 Pearson under *TERF1* CC ≤ -0.25 in high human left hemisphere compared with low chimpanzee left hemisphere (Fig. 1A-1D).

A

Ex-Inh-GRNInfer
ACOX3, ADD3, AF052141, AGL, AL042668, AL050030, ALG8, ARHGAP12, ATP5J2, BAP1, BCAS1, BLCAP, BTRC, C18orf10, C1orf61_1, CA2, CALM1, CAMTA1, CDR2, CFLAR, CGRRF1, CHAD, COL6A1, CTBP1, CYB5B, CYP2J2, DCI, DCTN1, DKFZp434H1419, DLEU1, DTNA, EPHX2, FDFT1, FKBPL, GDF10, GLOD4, GPD1, GPR68, GSTM3_2, GSTM5, GTF2A2, GTF2I_1, HABP4, HAX1, HCN2, HOMER1, HSPA1A, HSPA2, IDI1, IFI44L, ISCA1, ITPR1, JARID1D, KIAA0423, LGALS3BP, LIMCH1, LOC157627, LOC400642, LOH11CR2A, MAP1B_1, MAP1B_3, MEIS3P1, MFAP3L, MOAP1, MTUS1, NAIP, NCK2, NPAL3, NPTN, OPTN, PAK3, PCBP2_2, PCDHGA8, PCTK2, PDPN, PEPD, PKP4, POLR2J, PON2, PPID, PRDX6, PTTG1IP, RAB2A, RAB3GAP1, RAD50, RASSF2, RFK, RGL1, RNF14, RPA3, RPL23AP7, RPS26, SC4MOL, SELENBP1, SGSH, SH3BGR, SLC25A16, SLC25A46, STAT2, SULT1A2, SYNE2, TCF25, TUBGCP4_2, AB002448, AB016247, ABCB1, ACTR2, AL049242, AL049278, AL049987, ATP2B1, BTN3A3, CDC25B, CDS2, CLDN10, DDIT4, DDX3Y, DICER1, EGFR, ENPP2_1, ENPP2_2, EPM2AIP1, FABP6, FAM114A1, FCMD, FGF1, FLNB, FOXN3_1, FOXN3_2, GEM, GNAQ, H10776, HNRNPA1, HSD17B6, HYPE, INSIG1, ITGB3BP, KIAA0368, KIAA0644, KIAA0895, KIF3A, KRT10, LAMP1, LAMP2, MAF, MAP1B_2, MAPT, MED13, MED6, MGC15523, MIA3, N58318, NR1D2_1, NSUN5C, NUP214, OGG1_2, PALLD, PAX6, PCSK6, PDE4DIP, PDE8A, PDIA2, PHKG1, PMS2L1, PPARD, PRKCI_1, PRKRA, R37702, RAB7L1, RBBP6, RECQL, RFPL1S, ROM1, RP4_691N24.1, RPL13, SAPS2, SF3B3, SFPQ_2, SMA4, SMAD1_1, SSTR2, TH, THBD, TRAPPC6A, TULP3, UBR5, UNG, USP9Y, WDFY3, WDR68, WNK1, X81789, ZNF174, ZNF271, ZNF443, TFCP2, TIGR:, TJP2, TMEM63A, UPF3A, USP14, VDAC2, W26407, ZNF44

B

Ex-Inh-Pearson
ENPP2_1, ABCB1, ABTB2, ACOX3, ADD3, AF016004, AL031282, AL049987, AL109702, ALDH7A1, ALG8, ASAH1, BAP1, BCAS1, BEST1, C10orf10, C14orf1, C1orf61_1, C1orf61_2, C21orf33, CA2, CCRK, CD59, CDH19, CES2, CLDN10, COL6A2, COX8A, CTNNA1_1, CTNNA1_2, CYB5B, CYP2J2, DCI, DDIT4, DYNC2LI1, EGFR, ENOSF1, ENPP2_2, FAM114A1, FAM127A, FGF1, GAB2, GABRB1, GBE1, GEM, GLO1, GOLGA3, GSTM3_1, GSTM3_2, GSTM3_3, GSTM5, HABP4, HAX1, HLA_DPB1, HSD17B6, HSPA1A, HSPA2, IFI44L, INHBB, IPO13, JARID1D, KIAA0644, L20971, LAMP1, LAMP2, LGALS3BP, LILRA4, LOC730392, LOH11CR2A, LRRC42, MAF, MAN2B2, MARCKSL1, MFAP3L, MGC15523, MTMR15, NAIP, NIPA2, NPC1, NSUN5C, NUP214, OGFR, P4HB, PALLD, PCSK6, PDE8A, PDLIM5, PGD, PHKG1, PHLDB1, PMS2L4, PON2, PPP1R13B, PTTG1IP, RAC1, RASA4, RASSF2, RGL1, ROM1, RPL13, SELENBP1, SGSH, SH3BGR, SNRPE, SQSTM1, TCF25, TJP2, TMEM63A, TRAPPC6A, TYMS, UBB, UBR5, UNG, UTY, W28807, WASF3, WWOX, ZFP36L2

C

Ex-Inh-Overlap-GRNInfer and Pearson
ENPP2_1, ABCB1, ACOX3, ADD3, AL049987, ALG8, BAP1, BCAS1, C1orf61_1, CA2, CLDN10, CYB5B, CYP2J2, DCI, DDIT4, EGFR, ENPP2_2, FAM114A1, FGF1, GEM, GSTM3_2, GSTM5, HABP4, HAX1, HSD17B6, HSPA1A, HSPA2, IFI44L, JARID1D, KIAA0644, LAMP1, LAMP2, LGALS3BP, LOH11CR2A, MAF, MFAP3L, MGC15523, NAIP, NSUN5C, NUP214, PALLD, PCSK6, PDE8A, PHKG1, PON2, PTTG1IP, RASSF2, RGL1, ROM1, RPL13, SELENBP1, SGSH, SH3BGR, TCF25, TJP2, TMEM63A, TRAPPC6A, UBR5, UNG

D

Ex-Inh-Different Mutual Positive Pearson Correlation Compared with Con
ENPP2_1, FGF1, LGALS3BP, PON2, ADD3, LAMP1, MFAP3L, PDE8A, ROM1, TJP2, ALG8, CYP2J2, PTTG1IP, SELENBP1, TMEM63A, CA2, FAM114A1, GSTM3_2, HSPA1A, IFI44L, NUP214, C1orf61_1, JARID1D, RASSF2, NAIP, SGSH, RGL1

Figure 1 (A) *TERF1*-repressive molecules of high human left hemisphere by GRNInfer. (B) *TERF1*-repressive molecules of high human left hemisphere by Pearson. (C) *TERF1*-repressive overlapping molecules of high human left hemisphere by GRNInfer and Pearson. (D) *TERF1*-repressive different mutual-positive-correlation molecules in high human compared with the corresponding low chimpanzee left hemisphere. Con, chimpanzee left hemisphere; Ex, human left hemisphere; Inh, inhibition.

Materials and Methods

441 significant molecules were identified from 12558 genes of 14 high

human compared with 15 low chimpanzee left hemisphere in GEO data set GDS2678 (http://www.ncbi.nlm.nih.gov/sites/GDSbrowser?acc=GDS2678) containing brain cerebrum, anterior cingulated, anterior inferior parietal, anterior inferior temporal, middle frontal gyrus, frontal pole, etc. for studying high *TERF1* inside-outside-in inhibitive estrogen-induced morphogenesis through *LGALS3BP-SELENBP1-CA2-GSTM3* using significant analysis of microarrays (SAM) (http://www-stat.stanford.edu/~tibs/SAM/) [12]. The raw microarray data were processed by log base 2. Two classes were unpaired and minimum fold change $\geqslant 2$ selected (the false-discovery rate 0%).

Gene expression values of *TERF1-repressive* different molecules were computed in high human compared with the corresponding low chimpanzee left hemisphere by AVERAGE and STDEV.

TERF1-repressive different mutual-positive-correlation molecular Pearson coefficients were computed in high human left hemisphere compared with the corresponding low no-tumor hepatitis/cirrhotic tissues under *TERF1* CC $\leq$ -0.25, as measurements of the correlation (linear dependence) including two variables X and Y and giving an inclusive value between −1 and +1.

TERF1-repressive molecular network was further constructed in high human left hemisphere by GRNInfer [13], GVedit tool and our articles [14-31]. GRNInfer is a novel mathematic method called GNR (Gene Network Reconstruction tool) based on linear programming and a decomposition procedure for inferring gene networks. The method theoretically ensures the derivation of the most consistent network structure with respect to all of the datasets, thereby not only significantly alleviating the problem of data scarcity but also remarkably improving the reconstruction reliability. The following Equation (1) represents all of the possible networks for the same dataset. We established network based on the top 441 distinguished genes and selected parameters as lambda 0.0, threshold 1.0e-10.

$$J = (X' - A)U\Lambda^{-1}V^T + YV^T = \hat{J} + YV^T \tag{1}$$

TERF1-repressive molecular knowledge network was further calculated in high human left hemisphere based on terms and occurrence numbers of GO (Cellular Component, Molecular Function and Biological Process), KEGG, GenMAPP, BioCarta and Disease via Molecule Annotation System, MAS (CapitalBio Corporation, Beijing, China; http://bioinfo.capitalbio.com/mas3/). The primary databases of MAS integrated various well-known biological resources, such as Gene Ontology (http://www.geneontology.org), KEGG (http://www.genome.jp/kegg/), BioCarta (http://www.biocarta.com/), GenMapp (http://www.genmapp.org/), HPRD (http://www.hprd.org/), etc.

Result

TERF1-repressive different Pearson mutual-positive-correlation molecular gene expression values were illustrated column diagrams by AVERAGE and STDEV in high human and the corresponding low chimpanzee left hemisphere, as shown in Fig. 2A.

TERF1-repressive different mutual-positive-correlation molecular Pearson coefficients were illustrated column diagrams in high human compared with the corresponding low chimpanzee left hemisphere, as shown in Fig. 2B and 2C, respectively.

TERF1-repressive molecular network was further constructed by GRNInfer in high human left hemisphere, as shown in Fig. 3.

TERF1-repressive knowledge terms network was further identified by MAS 3.0 in high human left hemisphere, as shown in Fig. 4.

A

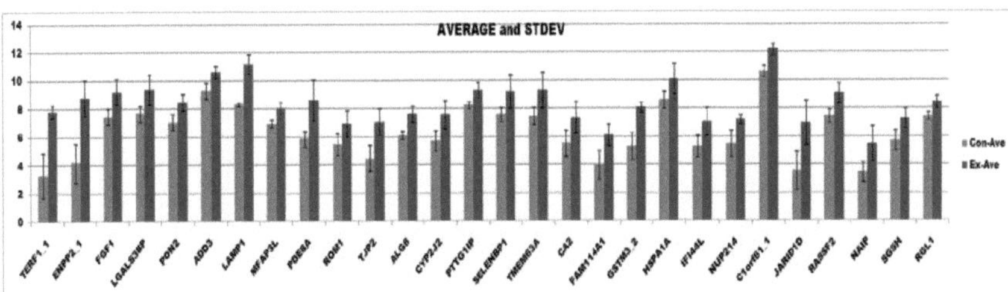

B

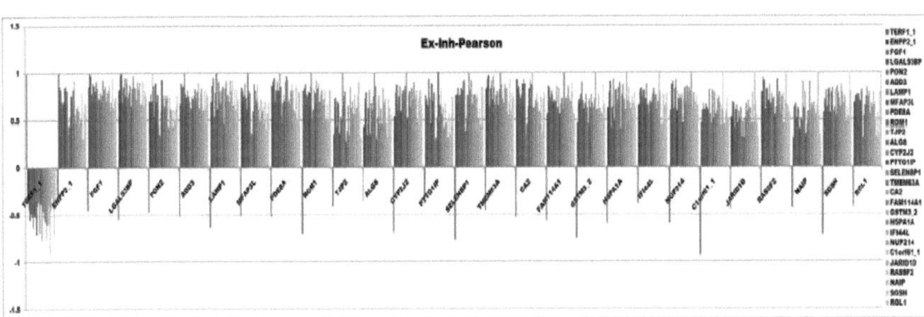

C

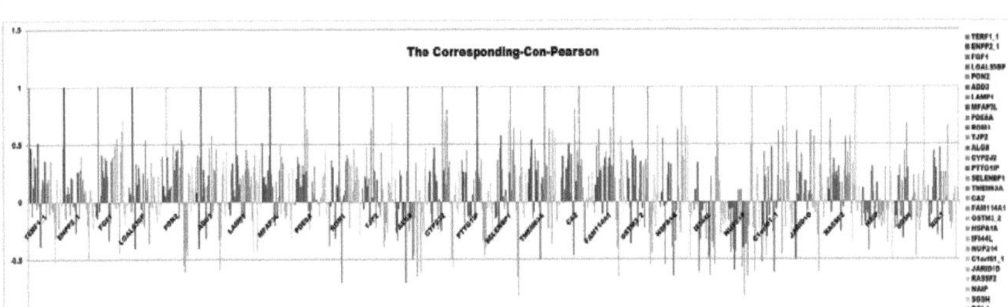

Figure 2 **(A)** Gene expression values of *TERF1*-repressive different mutual Pearson positive correlation molecules in high human and the corresponding low chimpanzee left hemisphere by AVERAGE and STDEV measurement. **(B)** Vertical quantification chart of *TERF1*-repressive different molecular mutual Pearson positive correlation coefficients in high human left hemisphere. n=14. **(C)** The corresponding correlation coefficients in low chimpanzee left hemisphere. n=15. Con, chimpanzee left hemisphere; Ex, human left hemisphere; Inh, inhibition.

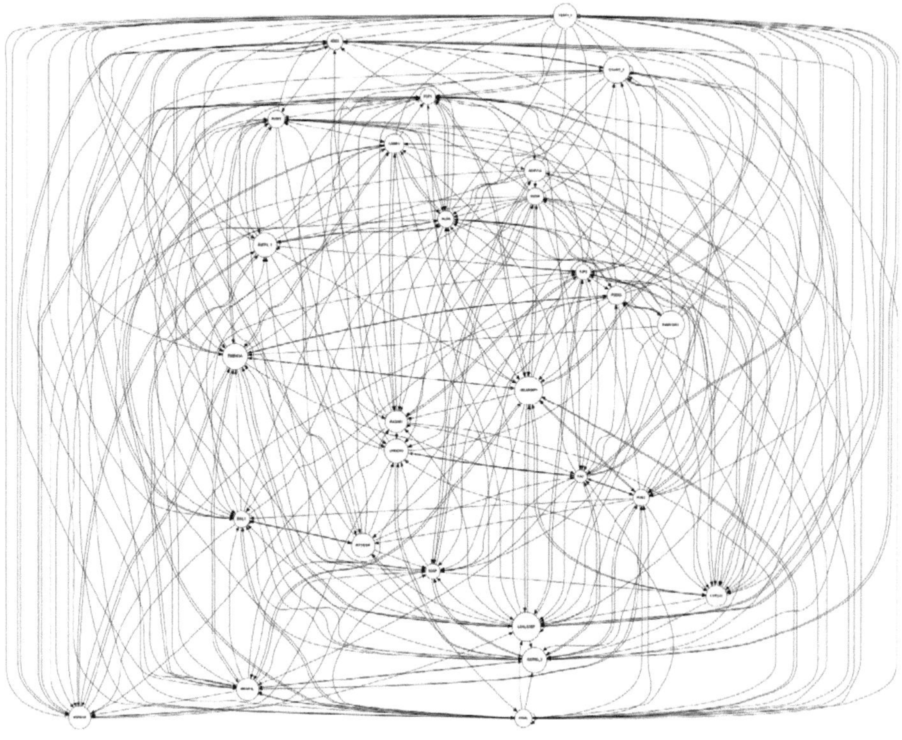

Figure 3 *TERF1*-repressive molecular network of high human left hemisphere by GRNInfer. n=14. Black arrow represents the activation relationship and empty circle as the inhibition one. Con, chimpanzee left hemisphere; Ex, human left hemisphere; Inh, inhibition.

Ex-Inh-Terms and Numbers

protein-binding	11	GTPase activator activity	2	myocardial ischemia	1	transferring hexosyl groups	1	unfolded protein-binding	1
nucleus	10	nucleic acid-binding	1	Sq- syndrome	1	protein amino acid N-linked glycosylation	1	mRNA catabolism	1
cytoplasm	10	phosphodiesterase I activity	1	cholera	1	N-Glycan biosynthesis	1	response to unfolded protein	1
membrane	7	nucleotide diphosphatase activity	1	solid tumor	1	glycoprotein metabolism	1	antigen processing and presentation	1
plasma membrane	6	alkylglycerophosphoethanolamine phosphodiesterase activity	1	malignant neoplasm of nasopharynx	1	protein amino acid glycosylation	1	HSP70 and apoptosis	1
metal ion-binding	5	phosphate metabolism	1	mesothelioma	1	glycoprotein biosynthesis	1	androgen receptor	1
signal transduction	5	G-protein coupled receptor protein signaling pathway	1	glioblastoma multiforme	1	biopolymer glycosylation	1	immune response	1
extracellular region	4	lipid catabolism	1	acute-phase reaction	1	congenital disorders of glycosylation overview	1	nuclear pore	1
hydrolase activity	4	ether lipid metabolism	1	muscular dystrophy limb girdle	1	arachidonic acid epoxygenase activity	1	cytosol	1
extracellular space	4	locomotion	1	lung metastasis	1	electron carrier activity	1	transporter activity	1
neoplasms	4	response to chemical substance	1	vascular proliferation	1	heme-binding	1	protein export from nucleus	1
malignant neoplasms	4	taxis	1	arteriosclerosis	1	icosanoid metabolism	1	mRNA transport	1
integral to membrane	4	locomotory behavior	1	cell transformation neoplastic	1	regulation of heart contraction	1	intracellular protein transmembrane transport	1
intracellular	4	heparin-binding	1	vitreoretinopathy proliferative	1	arachidonic acid metabolism	1	endomembrane system	1
integral to plasma membrane	3	angiogenesis	1	lymphatic vessel diseases	1	linoleic acid metabolism	1	TGF-beta receptor	1
metabolism	3	induction of an organ	1	bladder neoplasms	1	fatty acid metabolism	1	mechanism of protein import into the nucleus	1
extracellular matrix (sensu Metazoa)	3	development	1	fenestration	1	circulation	1	DNA-binding	1
MAPK signaling pathway	3	fibroblast growth factor receptor signaling	1	neoplasms epithelial	1	vesicular fraction	1	iron ion-binding	1
endocrine and CNS	3	morphogenesis	1	fibroadenoma	1	tryptophane metabolism	1	acting on single donors with incorporation of molecular oxygen	1
necrosis	3	cell differentiation	1	pathologic neovascularization	1	acting on paired donors	1	incorporation of two atoms of oxygen	1
adenocarcinoma	3	lung development	1	pancreatic carcinoma	1	with incorporation or reduction of molecular oxygen	1	chromatin modification	1
leukemia	3	positive regulation of epithelial cell proliferation	1	malignant mesothelioma	1	cytochrome P450	1	gametogenesis	1
lysosome	3	cell-cell signaling	1	neoplasms vascular tissue	1	eicosanoid metabolism	1	reproduction	1
cellular component	3	regulation of cell cycle	1	hiv encephalopathy	1	hepatitis B	1	sexual reproduction	1
endoplasmic reticulum	3	carbohydrate-binding	1	Alzheimer's disease	1	protein import into nucleus	1	male gamete generation	1
nucleotide-binding	3	BTG family proteins and cell cycle regulation	1	brain aneurysm	1	nuclear transport	1	acute GVHD	1
oxidoreductase activity	3	coronary occlusion	1	scavenger receptor activity	1	protein targeting	1	increased development of adult T-cell leukemia/lymphoma	1
transcription factor-binding	2	prostatic carcinoma	1	cellular defense response	1	nucleocytoplasmic transport	1	negative regulation of progression through cell	1
cell motility	2	neurilemmoma	1	aryldialkylphosphatase activity	1	pituitary neoplasms	1	blood and lymph	1
chemotaxis	2	prostatic hypertrophy benign	1	arylesterase activity	1	pituitary adenoma	1	nucleoside-triphosphatase activity	1
growth factor activity	2	loss of heterozygosity	1	identical protein-binding	1	selenium-binding	1	nervous system development	1
cell proliferation	2	adenocarcinoma of the stomach	1	aromatic compound catabolism	1	Selenoproteins	1	negative regulation of apoptosis	1
regulation of actin cytoskeleton	2	synovitis	1	gamma-hexachlorocyclohexane degradation	1	azoospermia | oligospermia	1	oncogenesis	1
melanoma	2	stage iv melanoma	1	bisphenol a degradation	1	carbonate dehydratase activity	1	negative regulation of programmed cell death	1
inflammation	2	achondroplasia	1	carboxylic ester hydrolase activity	1	morphogenesis of an epithelium	1	calcium ion-binding	1
pheochromocytoma	2	invasive cancer	1	actin-binding	1	one-carbon compound metabolism	1	sulfuric ester hydrolase activity	1
retinal diseases	2	senile plaques	1	essential hypertension	1	carbon dioxide transport	1	N-sulfoglucosamine sulfohydrolase activity	1
kaposi sarcoma	2	alzheimers disease	1	astrocytoma grade iii	1	secretion	1	proteoglycan metabolism	1
malignant neoplasm of breast	2	coronary atherosclerosis	1	arterial stiffness	1	nitrogen metabolism	1	glycosaminoglycan metabolism	1
glioma	2	vascular insufficiency of intestine	1	hypertension	1	hyperammonemia	1	glycosaminoglycan degradation	1
lymphoma	2	influenza	1	membrane fraction	1	metabolic diseases	1	multiple endocrine neoplasia type 2a	1
neoplasm metastasis	2	teratocarcinoma	1	lysosomal membrane	1	Vascular Disease	1	multiple endocrine neoplasia	1
rheumatoid arthritis	2	pancreatic neoplasms	1	multivesicular body	1	glutathione transferase activity	1	benign neoplasm	1
malignant glioma	2	ductal breast carcinoma	1	external side of plasma membrane	1	establishment of blood-nerve barrier	1	guanyl-nucleotide exchange factor activity	1
carcinoma squamous cell	2	inflammatory joint disease	1	endosome membrane	1	response to estrogen stimulus	1	Ral guanyl-nucleotide exchange factor activity	1
osteosarcoma	2	diabetic retinopathy	1	sarcolemma	1	glutathione metabolism	1	regulation of small GTPase mediated signal transduction	1
carcinoma	2	malignant neoplasm of endometrium	1	melanosome	1	metabolism of xenobiotics by cytochrome p450	1	small GTPase mediated signal transduction	1
astrocytoma	2	choroidal neovascularization	1	apparatus	1	drug metabolism - cytochrome p450	1	small GTPase regulator activity	1
glioblastoma	2	vesicular stomatitis	1	two-component response regulator activity	1	circadian exercise	1	enzyme activator activity	1
malignant neoplasm of pancreas	2	virus diseases	1	3',5'-cyclic-nucleotide phosphodiesterase activity	1	cardiomyopathy congestive	1	ENPP2_1 (ectonucleotide pyrophosphatase/phosphodiesterase 2)	
cancer of prostate	2	leukemia nonlymphocytic acute	1	manganese ion-binding	1	ventricular aneurysm	1	FGF1 (fibroblast growth factor 1 (acidic))	
retinal degeneration	2	chronic arthritis	1	two-component signal transduction system (phosphorelay)	1	myocardial infarction	1	LGALS3BP (lectin galactoside-binding soluble 3 binding protein)	
cell adhesion	2	amyotrophic lateral sclerosis	1	cAMP catabolism	1	acute myocardial infarction	1	PON2 (paraoxonase 2)	
cytoskeleton	2	bladder carcinoma	1	regulation of transcription DNA-dependent	1	skin cancer	1	ADD3 (adducin 3 (gamma))	
structural constituent of cytoskeleton	2	carcinoma transitional cell	1	cyclic nucleotide metabolism	1	non-melanoma	1	LAMP1 (lysosomal-associated membrane protein 1)	
calmodulin-binding	2	craniosynostosis	1	purine metabolism	1	head and neck cancer	1	MFAP3L (microfibrillar-associated protein 3-like)	
lytic vacuole	2	pathologic processes	1	G protein signaling	1	bladder cancer	1	PDE8A (phosphodiesterase 8A)	
vacuole	2	cancer of bladder	1	signal transducer activity	1	prostate cancer	1	ROM1 (retinal outer segment membrane protein 1)	
magnesium ion-binding	2	dysplasia	1	carcinoma of colon	1	liver disease	1	TJP2 (tight junction protein 2 (zona occludens 2))	
malignant neoplasm of lung	2	chondrosarcoma	1	adenoma	1	alcoholic	1	ALG8 (asparagine-linked glycosylation 8 alpha-1,3-glucosyltransferase homolog (S. cerevisiae))	
visual perception	2	carcinoma papillary	1	adenocarcinoma of the prostate	1	oral cancer | leukoplakia	1	CYP2J2 (cytochrome P450 family 2 subfamily J polypeptide	
retinitis pigmentosa	2	uterine fibroids	1	cholangiocarcinoma	1	glaucoma	1	PTTG1IP (pituitary tumor-transforming 1 interacting protein)	
Asthma	2	diphtheria	1	primary carcinoma of the liver cells	1	primary open-angle	1	SELENBP1 (selenium binding protein 1)	
tight junction	2	ischemia	1	malignant tumor of colon	1	ototoxicity	1	TMEM63A (transmembrane protein 63A)	
cell junction	2	adenoma pleomorphic	1	adenovirus infections	1	laryngeal cancer	1	CA2 (carbonic anhydrase II)	
endoplasmic reticulum membrane	2	aggressiveness	1	brain neoplasms	1	acrylonitrile metabolism/toxicity | ethylene oxide metabolism/toxicity	1	FAM114A1 (family with sequence similarity 114 member A1)	
transferase activity	2	[illegible] carcinoma	1	ovarian cancer	1	pancreatitis	1	GSTM3_2 (glutathione S-transferase mu 3 (brain))	
microsome	2	neurodegenerative diseases	1	chronic pancreatitis	1	stomach cancer	1	HSPA1A (heat shock 70kDa protein 1A)	
oxidation reduction	2	fibrosarcoma	1	response to stimulus	1	oral cancer	1	IFI44L (interferon-induced protein 44-like)	
molecular function	2	osteoarthropathy	1	sensory perception of light	1	cystic fibrosis	1	NUP214 (nucleoporin 214kDa)	
protein transport	2	hyperplasia	1	atrophy	1	breast cancer	1	C1orf61_1 (chromosome 1 open reading frame 61)	
zinc ion-binding	2	autoimmune diseases	1	meningococcal infections	1	alcohol abuse | smoking behavior	1	JARID1D (jumonji AT rich interactive domain 1D)	
lyase activity	2	adenoma follicular	1	macular dystrophy	1	colorectal cancer	1	RASSF2 (Ras association (RalGDS/AF-6) domain family member 2)	
anti-apoptosis	2	barrett esophagus	1	vitelliform	1	lung cancer	1	NAIP (NLR family apoptosis inhibitory protein)	
spermatogenesis	2	carcinoma renal cell	1	guanylate kinase activity	1	lung cancer | hamartomas	1	SGSH (N-sulfoglucosamine sulfohydrolase)	
apoptosis	2	monosomy	1	cholera - infection	1	mitochondrion	1	RGL1 (ral guanine nucleotide dissociation stimulator-like 1)	
neurofibromatosis 1	2	myelodysplastic syndromes	1	alpha-1,3-mannosyltransferase activity	1	ATP-binding	1		

Figure 4 *TERF1*-repressive terms and occurrence numbers in high human left hemisphere based on GO, KEGG, GenMAPP, BioCarta and Disease via MAS 3.0. Ex, human left hemisphere; Inh, inhibition.

Discussion

TERF1-repressive different molecular Pearson mutual-positive-correlation network was setup in high human compared with the corresponding low chimpanzee left hemisphere (Fig. 3). Our identified

TERF1 inside-outside-in inhibitive molecular network showed *LGALS3BP* (lectin galactoside-binding soluble 3-binding protein), *SELENBP1* (selenium-binding protein 1), *CA2* (carbonic anhydrase II), *GSTM3* (glutathione S-transferase mu 3 (brain)) in high human left hemisphere. *TERF1* inside-outside-in inhibitive terms network includes extracellular region, extracellular matrix (sensu Metazoa), extracellular space, nucleus, membrane, cytoplasm; cellular defense response, response to estrogen stimulus; cell adhesion, morphogenesis of an epithelium; scavenger receptor activity, protein-binding, selenium-binding, carbonate dehydratase activity, zinc ion-binding, lyase activity, metal ion-binding, glutathione transferase activity, transferase activity in high human left hemisphere based on integrative GO, KEGG, GenMAPP, BioCarta and disease databases (Fig. 4). Therefore, we propose high *TERF1* inside-outside-in inhibitive estrogen-induced morphogenesis through *LGALS3BP-SELENBP1-CA2-GSTM3* in human left hemisphere.

Morphogenesis positive relationship with lectin, carbonic, glutathione has been reported in references as follows: Basal laminar thinning in branching morphogenesis of the chick lung as demonstrated by lectin probes; Lectin histochemistry in normal and abnormal limb morphogenesis in the mouse; Localization of endogenous galactoside-binding lectin during morphogenesis of Xenopus laevis [32-34]. A role for carbonic anhydrase in early eye morphogenesis [35]. Role of glutathione in the morphogenesis of the bacterial spore coat; An interaction between glutathione and the capsid is required for the morphogenesis of C-cluster enteroviruses; Binding of glutathione to enterovirus capsids is essential for virion morphogenesis [36-38].

In summary, *TERF1*-repressive molecular Pearson mutual-positive-correlation network was constructed in high human left hemisphere from the overlapping molecules of GRNInfer with Pearson. We propose and verify high *TERF1* inside-outside-in inhibitive estrogen-induced morphogenesis through *LGALS3BP-SELENBP1-CA2-GSTM3* in human left hemisphere. New critical insights and perspectives for the future will be further verified our hypothesis by experimental biology. High *TERF1* inside-outside-in inhibitive estrogen-induced morphogenesis through *LGALS3BP-SELENBP1-CA2-GSTM3* is very useful to identify novel markers and potential drugs for prognosis and therapy, develop a new route for studying the pathogenesis.

References

[1] Retraction. Bone morphogenetic protein-7 induces telomerase inhibition, telomere shortening, breast cancer cell senescence, and death via Smad3. FASEB J. [J] 23(8). 2009. 2790.

[2] Cassar, L., H. Li, A.R. Pinto, et al. Bone morphogenetic protein-7 inhibits telomerase activity, telomere maintenance, and cervical tumor growth. Cancer Res. [J] 68(22). 2008. 9157-66.

[3] Cassar, L., C. Nicholls, A.R. Pinto, et al. Bone morphogenetic protein-7 induces telomerase inhibition, telomere shortening, breast cancer cell senescence, and death via Smad3. FASEB J. [J] 23(6). 2009. 1880-92.

[4] Camacho Leal Mdel, P., A. Pincini, G. Tornillo, et al. p130Cas over-expression impairs mammary branching morphogenesis in response to estrogen and EGF. PLoS One. [J] 7(12). 2012. e49817.

[5] Fan, X., H.J. Kim, M. Warner, et al. Estrogen receptor beta is essential for sprouting of nociceptive primary afferents and for morphogenesis and maintenance of the dorsal horn interneurons. Proc Natl Acad Sci U S A. [J] 104(34). 2007. 13696-701.

[6] Feng, Y., D. Manka, K.U. Wagner, et al. Estrogen receptor-alpha expression in the mammary epithelium is required for ductal and alveolar morphogenesis in mice. Proc Natl Acad Sci U S A. [J] 104(37). 2007. 14718-23.

[7] Mallepell, S., A. Krust, P. Chambon, et al. Paracrine signaling through the epithelial estrogen receptor alpha is required for proliferation and morphogenesis in the mammary gland. Proc Natl Acad Sci U S A. [J] 103(7). 2006. 2196-201.

[8] Park, S., Y. Zhao, S. Yoon, et al. Repressor of estrogen receptor activity (REA) is essential for mammary gland morphogenesis and functional activities: studies in conditional knockout mice. Endocrinology. [J] 152(11). 2011. 4336-49.

[9] Rossetti, S., F. Corlazzoli, A. Gregorski, et al. Identification of an estrogen-regulated circadian mechanism necessary for breast acinar morphogenesis. Cell Cycle. [J] 11(19). 2012. 3691-700.

[10] Shekhar, M.P., J. Werdell, and L. Tait. Interaction with endothelial cells is a prerequisite for branching ductal-alveolar morphogenesis and hyperplasia of preneoplastic human breast epithelial cells: regulation by estrogen. Cancer Res. [J] 60(2). 2000. 439-49.

[11] Xu, X., Y. Lu, G. Zhang, et al. Bisphenol A promotes dendritic morphogenesis of hippocampal neurons through estrogen receptor-mediated ERK1/2 signal pathway. Chemosphere. [J] 96. 2014. 129-37.

[12] Storey., J.D. A direct approach to false discovery rates. J. Roy. Stat. Soc., Ser. B. [J] 64. 2002. 479–498.

[13] Wang, Y., T. Joshi, X.S. Zhang, et al. Inferring gene regulatory networks from multiple microarray datasets. Bioinformatics. [J] 22(19). 2006. 2413-20.

[14] Wang, L., J. Huang, M. Jiang, et al. Activated PTHLH Coupling Feedback Phosphoinositide to G-Protein Receptor Signal-Induced Cell Adhesion Network in Human Hepatocellular Carcinoma by Systems-Theoretic Analysis. ScientificWorldJournal. [J] 2012. 2012. 428979.

[15] Wang, L., J. Huang, M. Jiang, et al. Inhibited PTHLH downstream leukocyte adhesion-mediated protein amino acid N-linked glycosylation coupling Notch and JAK-STAT cascade to iron-sulfur cluster assembly-induced aging network in no-tumor hepatitis/cirrhotic tissues (HBV or HCV infection) by systems-theoretical analysis. Integr Biol (Camb). [J] 4(10). 2012. 1256-62.

[16] Wang, L., J. Huang, M. Jiang, et al. Tissue-specific transplantation antigen P35B (TSTA3) immune response-mediated metabolism coupling cell cycle to postreplication repair network in no-tumor hepatitis/cirrhotic tissues (HBV or HCV infection) by biocomputation. Immunol Res. [J] 52(3). 2012. 258-68.

[17] Wang, L., J. Huang, M. Jiang, et al. Signal transducer and activator of transcription 2 (STAT2) metabolism coupling postmitotic outgrowth to visual and sound perception network in human left cerebrum by biocomputation. J Mol Neurosci. [J] 47(3). 2012. 649-58.

[18] Lin, H., L. Wang, M. Jiang, et al. P-glycoprotein (ABCB1) inhibited network of mitochondrion transport along microtubule and BMP signal-induced cell shape in chimpanzee left cerebrum by systems-theoretical analysis. Cell Biochem Funct. [J] 30(7). 2012. 582-7.

[19] Huang, J., L. Wang, M. Jiang, et al. PTHLH coupling upstream negative regulation of fatty acid biosynthesis and Wnt receptor signal to downstream peptidase activity-induced apoptosis network in human hepatocellular carcinoma by systems-theoretical analysis. J Recept Signal Transduct Res. [J] 32(5). 2012. 250-6.

[20] Wang, L., L. Sun, J. Huang, et al. Cyclin-dependent kinase inhibitor 3 (CDKN3) novel cell cycle computational network between human non-malignancy associated hepatitis/cirrhosis and hepatocellular carcinoma (HCC) transformation. CELL PROLIFERAT. [J] 44(3). 2011. 291-9.

[21] Wang, L., J. Huang, M. Jiang, et al. AFP computational secreted network construction and analysis between human hepatocellular carcinoma (HCC) and no-tumor hepatitis/cirrhotic liver tissues. Tumour Biol. [J] 31(5). 2011. 417-25.

[22] Wang, L., J. Huang, M. Jiang, et al. Survivin (BIRC5) cell cycle computational network in human no-tumor hepatitis/cirrhosis and hepatocellular carcinoma transformation. J Cell Biochem. [J] 112(5). 2011. 1286-94.

[23] Wang, L., J. Huang, M. Jiang, et al. MYBPC1 computational phosphoprotein network construction and analysis between frontal cortex of HIV encephalitis (HIVE) and HIVE-control patients. Cell Mol Neurobiol. [J] 31(2). 2011. 233-41.

[24] Wang, L., J. Huang, and M. Jiang. CREB5 computational regulation network construction and analysis between frontal cortex of HIV encephalitis (HIVE) and HIVE-control patients. Cell Biochem Biophys. [J] 60(3). 2011. 199-207.

[25] Wang, L., J. Huang, and M. Jiang. RRM2 computational phosphoprotein network construction and analysis between no-tumor hepatitis/cirrhotic liver tissues and human hepatocellular carcinoma (HCC). Cell Physiol Biochem. [J] 26(3). 2011. 303-10.

[26] Sun, L., L. Wang, M. Jiang, et al. Glycogen debranching enzyme 6 (*AGL*), enolase 1 (ENOSF1), ectonucleotide pyrophosphatase 2 (ENPP2_1), glutathione S-transferase 3 (*GSTM3*_3) and mannosidase (MAN2B2) metabolism computational network analysis between chimpanzee and human left cerebrum. Cell Biochem Biophys. [J] 61(3). 2011. 493-505.

[27] Huang, J.X., L. Wang, and M.H. Jiang. TNFRSF11B computational development network construction and analysis between frontal cortex of HIV encephalitis (HIVE) and HIVE-control patients. J Inflamm (Lond). [J] 7. 2011. 50.

[28] Sun, Y., L. Wang, M. Jiang, et al. Secreted Phosphoprotein 1 Upstream Invasive Network Construction and Analysis of Lung Adenocarcinoma Compared with Human Normal Adjacent

Tissues by Integrative Biocomputation. Cell Biochem Biophys. [J] 56(2-3). 2010. 59-71.

[29] Huang, J., L. Wang, M. Jiang, et al. Interferon α-Inducible Protein 27 Computational Network Construction and Comparison between the Frontal Cortex of HIV Encephalitis (HIVE) and HIVE-Control Patients The Open Genomics Journal [J] 3(1875-693X). 2010. 1-8.

[30] Wang, L., Y. Sun, M. Jiang, et al. Integrative decomposition procedure and Kappa statistics for the distinguished single molecular network construction and analysis. J Biomed Biotechnol. [J] 2009. 2009. 726728.

[31] Wang, L., Y. Sun, M. Jiang, et al. FOS proliferating network construction in early colorectal cancer (CRC) based on integrative significant function cluster and inferring analysis. Cancer Invest. [J] 27(8). 2009. 816-24.

[32] Gallagher, B.C. Basal laminar thinning in branching morphogenesis of the chick lung as demonstrated by lectin probes. J Embryol Exp Morphol. [J] 94. 1986. 173-88.

[33] Milaire, J. Lectin histochemistry in normal and abnormal limb morphogenesis in the mouse. Prog Histochem Cytochem. [J] 23(1-4). 1991. 132-40.

[34] Milos, N.C., Y.L. Ma, P.V. Varma, et al. Localization of endogenous galactoside-binding lectin during morphogenesis of Xenopus laevis. Anat Embryol (Berl). [J] 182(4). 1990. 319-27.

[35] Linser, P.J. and J.A. Plunkett. A role for carbonic anhydrase in early eye morphogenesis. Invest Ophthalmol Vis Sci. [J] 30(4). 1989. 783-5.

[36] Cheng, H.M., A.I. Aronson, and S.C. Holt. Role of glutathione in the morphogenesis of the bacterial spore coat. J Bacteriol. [J] 113(3). 1973. 1134-43.

[37] Ma, H.C., Y. Liu, C. Wang, et al. An interaction between glutathione and the capsid is required for the morphogenesis of C-cluster enteroviruses. PLoS Pathog. [J] 10(4). 2014. e1004052.

[38] Thibaut, H.J., L. van der Linden, P. Jiang, et al. Binding of glutathione to enterovirus capsids is essential for virion morphogenesis. PLoS Pathog. [J] 10(4). 2014. e1004039.

Chapter 2: High *USP14* feedback-inhibitive axon, T cell induced microtubule complex through *MAPT-RASGRP1-RFK-AF052119-M19267*

Abstract

29 different Pearson mutual-positive-correlation *USP14*-repressive molecular network was constructed from 46 overlapping of 148 GRNInfer and 137 Pearson under *USP14* CC ≤ -0.25 in high human left hemisphere compared with low chimpanzee left hemisphere. Our identified *USP14* feedback-inhibitive molecular network showed *MAPT* (microtubule-associated protein tau), *RASGRP1* (RAS guanyl releasing protein 1 (calcium and DAG-regulated)), *RFK* (riboflavin kinase), *AF052119* (solute carrier family 25 (mitochondrial carrier; adenine nucleotide translocator) member 4), *M19267* (Human tropomyosin mRNA complete cds) in high human left hemisphere. *USP14* feedback-inhibitive terms network includes cytoplasm, cytosol, plasma membrane, cell projection, membrane, intracellular, membrane fraction, endoplasmic reticulum, endoplasmic reticulum membrane; axon, positive regulation of axon extension, generation of neurons, endocrine and CNS, dementia | neuropathy, T cell receptor signaling pathway; microtubule, microtubule associated complex, microtubule-binding, microtubule cytoskeleton organization and biogenesis, negative regulation of microtubule depolymerization, positive regulation of microtubule polymerization, microtubule cytoskeleton, apoptosis, microtubule-based process, cell differentiation; lipoprotein-binding, SH3 domain-binding, enzyme-binding, identical protein-binding, apolipoprotein

E-binding, guanyl-nucleotide exchange factor activity, calcium ion-binding, protein-binding, zinc ion-binding, diacylglycerol-binding, nucleotide-binding, magnesium ion-binding, ATP-binding, riboflavin kinase activity, transferase activity based on integrative GO, KEGG, GenMAPP, BioCarta and disease databases in high human left hemisphere. Therefore, we propose high *USP14* feedback-inhibitive axon, T cell induced microtubule complex through *MAPT-RASGRP1-RFK-AF052119-M19267* in human left hemisphere.

Keywords: *USP14* feedback-inhibitive network; microtubule complex; axon; T cell

Introduction

USP14 is one of our identified significant molecules (fold change 2) in high human left hemisphere compared with the corresponding low chimpanzee left hemisphere. *USP14* appears cytoplasm; cysteine-type endopeptidase activity, ubiquitin thiolesterase activity, ubiquitin-specific protease activity, tRNA guanylyltransferase activity, peptidase activity; protein modification, ubiquitin-dependent protein catabolism; ubiquitin-dependent protein catabolism, cysteine-type peptidase activity, ubiquitin cycle, cysteine-type endopeptidase activity, nucleotidyltransferase activity based on GO, KEGG, GenMAPP, BioCarta and disease databases. Ubiquitin negative relationship with microtubule has been reported in references as follows: Axotrophin/MARCH7 acts as an E3 ubiquitin ligase and ubiquitinates tau protein in vitro impairing microtubule binding; Levels of the ubiquitin ligase substrate adaptor MEL-26 are inversely correlated with MEI-1/katanin microtubule-severing activity during both meiosis and mitosis; The C. elegans anaphase promoting complex and MBK-2/DYRK kinase act redundantly with CUL-3/MEL-26 ubiquitin ligase to degrade MEI-1 microtubule-severing activity after meiosis [1-3]. Axon and T cell relation with microtubule has been reported in references as follows: The Microtubule Minus-End-Binding Protein Patronin/PTRN-1 Is Required for Axon Regeneration in C. elegans; Microtubule dynamics in axon guidance; Neuronal deletion of GSK3beta increases microtubule speed in the growth cone and enhances axon regeneration via CRMP-2 and independently of MAP1B and CLASP2;

TACC3 is a microtubule plus end-tracking protein that promotes axon elongation and also regulates microtubule plus end dynamics in multiple embryonic cell types [4-7]. An experimental and computational study of effects of microtubule stabilization on T-cell polarity; Mammalian diaphanous-related formin 1 regulates GSK3beta-dependent microtubule dynamics required for T cell migratory polarization; The polarity protein Par1b/EMK/MARK2 regulates T cell receptor-induced microtubule-organizing center polarization [8-10]. Yet high *USP14* feedback-inhibitive axon, T cell induced microtubule complex through *MAPT-RASGRP1-RFK-AF052119-M19267* in human left hemisphere is not clear.

29 different Pearson mutual-positive-correlation *USP14*-repressive molecular network was constructed from 46 overlapping of 148 GRNInfer and 137 Pearson under *USP14* CC ≤ -0.25 in high human left hemisphere compared with low chimpanzee left hemisphere (Fig. 5A-5D).

A

Ex-Inh-GRNInfer
TUBGCP4_2, ABCC10, AF052119, AF052141, AL042668, AL049242, AL049278, ALG8, BAP1, BCAS1, BTRC, C1orf61_1, CCNO, CDH19, CDR2, CGRRF1, CHAD, CLDN10, COL6A1, CYB5B, DDX19A, DDX3Y, DIXDC1, DTNA, DYNC1I1, DYNC2LI1, EVI5, FBXL5, FCMD, FDFT1, FLJ43806, FOXN3_2, GAB2, GDF10, GIPC3, GPD1, GSTM3_2, GSTM5, GTF2A2, GUSBP1, HABP4, HCN2, HNRPH3, HSD17B6, KIAA0888, LGALS3BP, LOC157627, LOH11CR2A, M19267, MAP1B_1, MAP1B_3, MAPT, MEIS3P1, MGC15523, NAIP, NCK2, NSUN5C, OGG1_1, OSBPL8, PAK3, PCSK6, PCTK2, PDPN, PEPD, PGD, PIN1, POLR2J, PON2, PPP1CA, PPP1R13B, PRKCI_1, PRKDC, PTTG1IP, RAB2A, RAPGEF4, RASGRP1, RBBP6, RBCK1, RFK, RFPL1S, RNF14, RP4_691N24.1, RPA3, RPL23AP7, SC4MOL, SEL1L, SELENBP1, SFRS7, SH3BGR, SLC25A46, SMAD1_1, SMAD1_2, SMC5, SMG1, SRI, STAT2, SULT1A2, TFCP2, THBS2, TIGR:, TLOC1, TMED10, TRAPPC6A, TULP3, ABCB1, ACADM, AF016004, AGL, AMD1, BMP2K, CFHR1, CFLAR, CTRL, CYFIP2, EIF1AY, ENOSF1, FOXN3_1, GATAD1, HSP90AB1, HSPA2, ITPR1, KIAA1109, LOC730392, MAGOH, MAP1B_2, MED13, NR1D2_2, PALM2_AKAP2, PDE8A, PKP4, PPARD, PROSC, R37702, RCBTB2, SARM1, SCYE1, SFRS1, SLC25A6, SMA4, SYNE2, TERF1_1, USP46, W22289, WIPF2, X82895, ZNF443, WNK1, ZNF271

B

Ex-Inh-Pearson
ELNR1, TUBGCP4_1, TUBGCP4_2, AA975427, ACADM, ACTR2, AF052119, AGL, AK3P1, AL050030, AL080234, AL109696, ANAPC10, ARF6, ATP2B1, ATP5J, ATP5J2, AW043812, BRP44, BTRC, C1D, CACNB3, CALM1, CAMTA1, CCNO, CDKN1B, CDR2, CGRRF1, CLASP2, CLCN4, CTBP1, DDX19A, DGCR5, DICER1, DKFZp434H1419, DLEU1, DTNA, EGFR, EID1, EXTL2, FBXL5, FEZ1, FOXN3_2, GIPC3, GPR68, GTF2A2, GTF2I_1, H10776, H2AFX, HCN2, HNRNPA0, HOMER1, HSPA9, HYPE, IQCK, ISCA1, ITGB3BP, ITPR1, KCNK1, KIAA0368, KIAA0423, KIAA0888, KIAA0895, KIAA1109, KIF3A, KPNB1, LOH11CR2A, M19267, MAP1B_1, MAP1B_2, MAPT, MEIS3P1, MIA3, MOAP1, MTHFS, NDRG4, NDUFA5, NFIB, NPAL3, NPTN, NR1D2_1, NR1D2_2, OSBPL8, PCDHGA8, PCTK2, PDE4DIP, PER2, PIN1, POLR2J, PPP1CA, PRDX2, PRKCI_1, PRKCI_2, PRKDC, PRKRA, PRPF19, PSMA4, R37702, RAB3GAP1, RAD50, RASGRP1, RBCK1, RCHY1, RFK, RFPL1S, RNF2, RPA3, RPL23AP7, RPS26, SEL1L, SERINC3, SF3B3, SFRS1, SFRS7, SLC25A6, SLC35E2, SMG1, SNW1, SPAG9, SPTB, SSTR2, TERF1_2, TH, THBD, TIPRL, TSPAN5, TULP3, U00928, U59632, USP8, VDAC2, W26407, WDFY3, WDR68, WIPF2, ZNF91_1, ZNF91_2

C

Ex-Inh-Overlap-GRNInfer and Pearson
TUBGCP4_2, ACADM, AF052119, AGL, BTRC, CCNO, CDR2, CGRRF1, DDX19A, DTNA, FBXL5, FOXN3_2, GIPC3, GTF2A2, HCN2, ITPR1, KIAA0888, KIAA1109, LOH11CR2A, M19267, MAP1B_1, MAP1B_2, MAPT, MEIS3P1, NR1D2_2, OSBPL8, PCTK2, PIN1, POLR2J, PPP1CA, PRKCI_1, PRKDC, R37702, RASGRP1, RBCK1, RFK, RFPL1S, RPA3, RPL23AP7, SEL1L, SFRS1, SFRS7, SLC25A6, SMG1, TULP3, WIPF2

D

Ex-Inh-Different Mutual Positive Pearson Correlation Compared with Con
MAPT, RASGRP1, TULP3, DDX19A, ITPR1, KIAA1109, PRKCI_1, SEL1L, BTRC, CDR2, OSBPL8, RFK, SMG1, SFRS1, MAP1B_2, CGRRF1, GTF2A2, MEIS3P1, NR1D2_2, POLR2J, PRKDC, SFRS7, FBXL5, PCTK2, TUBGCP4_2, AF052119, KIAA0888, M19267, R37702

Figure 5 (A) *USP14*-repressive molecules of high human left hemisphere by GRNInfer. **(B)** *USP14*-repressive molecules of high human left hemisphere by Pearson. **(C)** *USP14*-repressive overlapping molecules of high human left hemisphere by GRNInfer and Pearson. **(D)** *USP14*-repressive different mutual-positive-correlation molecules in high human left hemisphere compared with the corresponding low chimpanzee left hemisphere. Con, chimpanzee left hemisphere; Ex, human left hemisphere; Inh, inhibition.

Materials and Methods

441 significant molecules were identified from 12558 genes of 14 high human compared with 15 low chimpanzee left hemisphere in GEO data set GDS2678 (http://www.ncbi.nlm.nih.gov/sites/GDSbrowser?acc=GDS2678) containing brain cerebrum, anterior cingulated, anterior inferior parietal, anterior inferior temporal, middle frontal gyrus, frontal pole, etc. for studying high *USP14* feedback-inhibitive axon, T cell induced microtubule complex through *MAPT-RASGRP1-RFK-AF052119-M19267* using significant analysis of microarrays (SAM) (http://www-stat.stanford.edu/~tibs/SAM/) [11]. The raw microarray data were processed by log base 2. Two classes were unpaired and minimum fold change ⩾2 selected (the false-discovery rate 0%).

Gene expression values of *USP14-repressive* different molecules were computed in high human left hemisphere compared with the corresponding low chimpanzee left hemisphere by AVERAGE and STDEV.

USP14-repressive different mutual-positive-correlation molecular Pearson coefficients were computed in high human left hemisphere compared with the corresponding low no-tumor hepatitis/cirrhotic tissues under *USP14* CC ≤ -0.25, as measurements of the correlation (linear dependence) including two variables X and Y and giving an inclusive value between −1 and +1.

USP14-repressive molecular network was further constructed in high human left hemisphere by GRNInfer [12], GVedit tool and our articles [13-30]. GRNInfer is a novel mathematic method called GNR (Gene Network Reconstruction tool) based on linear programming and a decomposition procedure for inferring gene networks. The method theoretically ensures the derivation of the most consistent network structure with respect to all of the datasets, thereby not only significantly alleviating the problem of data scarcity but also remarkably improving the reconstruction reliability. The following Equation (1) represents all of the possible networks for the same dataset. We established network based on the top 441 distinguished genes and selected parameters as lambda 0.0, threshold 1.0e-10.

$$J = (X' - A)U\Lambda^{-1}V^{T} + YV^{T} = \hat{J} + YV^{T} \quad (1)$$

USP14-repressive molecular knowledge network was further calculated in high human left hemisphere based on terms and occurrence numbers of GO (Cellular Component, Molecular Function and Biological Process), KEGG, GenMAPP, BioCarta and Disease via Molecule Annotation System, MAS (CapitalBio Corporation, Beijing, China; http://bioinfo.capitalbio.com/mas3/). The primary databases of MAS integrated various well-known biological resources, such as Gene Ontology (http://www.geneontology.org), KEGG (http://www.genome.jp/kegg/), BioCarta (http://www.biocarta.com/), GenMapp (http://www.genmapp.org/), HPRD (http://www.hprd.org/), etc.

Result

USP14-repressive different Pearson mutual-positive-correlation molecular gene expression values were illustrated column diagrams by AVERAGE and STDEV in high human left hemisphere and the corresponding low chimpanzee left hemisphere, as shown in Fig. 6A.

USP14-repressive different mutual-positive-correlation molecular

Pearson coefficients were illustrated column diagrams in high human left hemisphere compared with the corresponding low chimpanzee left hemisphere, as shown in Fig. 6B and 6C, respectively.

USP14-repressive molecular network was further constructed by GRNInfer in high human left hemisphere, as shown in Fig. 7.

USP14-repressive knowledge terms network was further identified by MAS 3.0 in high human left hemisphere, as shown in Fig. 8.

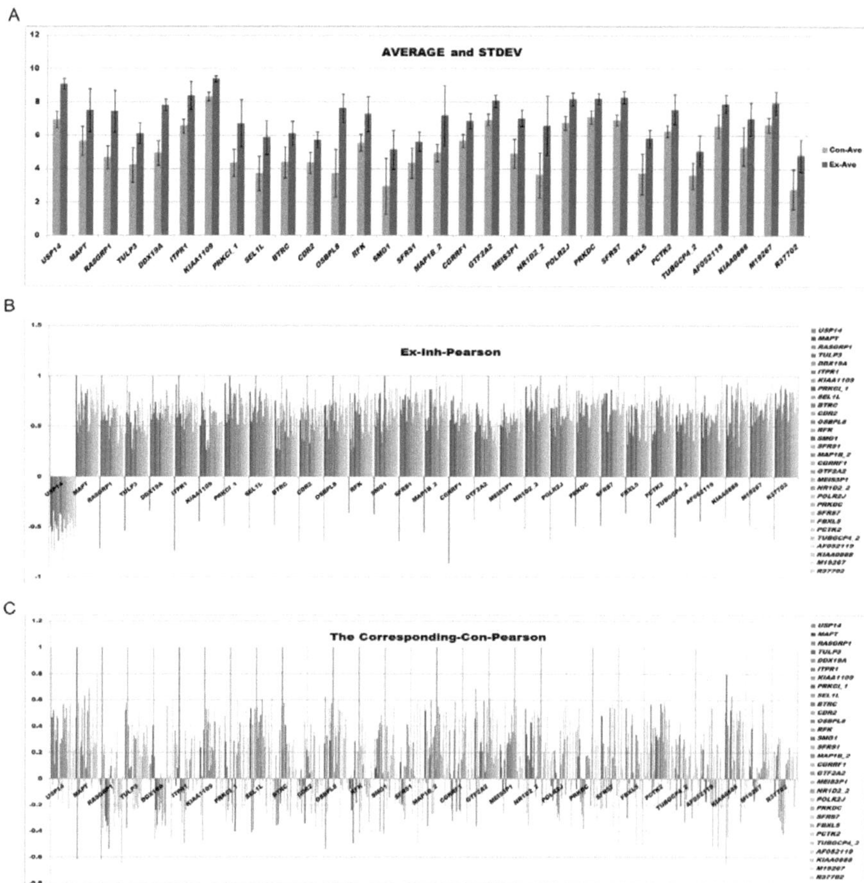

Figure 6 (A) Gene expression values of *USP14*-repressive different mutual Pearson positive correlation molecules in high human left hemisphere and the corresponding low chimpanzee left hemisphere by AVERAGE and STDEV measurement. **(B)** Vertical quantification chart of *USP14*-repressive different molecular mutual Pearson positive correlation coefficients in high human left hemisphere. n=14. **(C)** The corresponding correlation coefficients in low chimpanzee left hemisphere. n=15. Con, chimpanzee left hemisphere; Ex, human left hemisphere; Inh, inhibition.

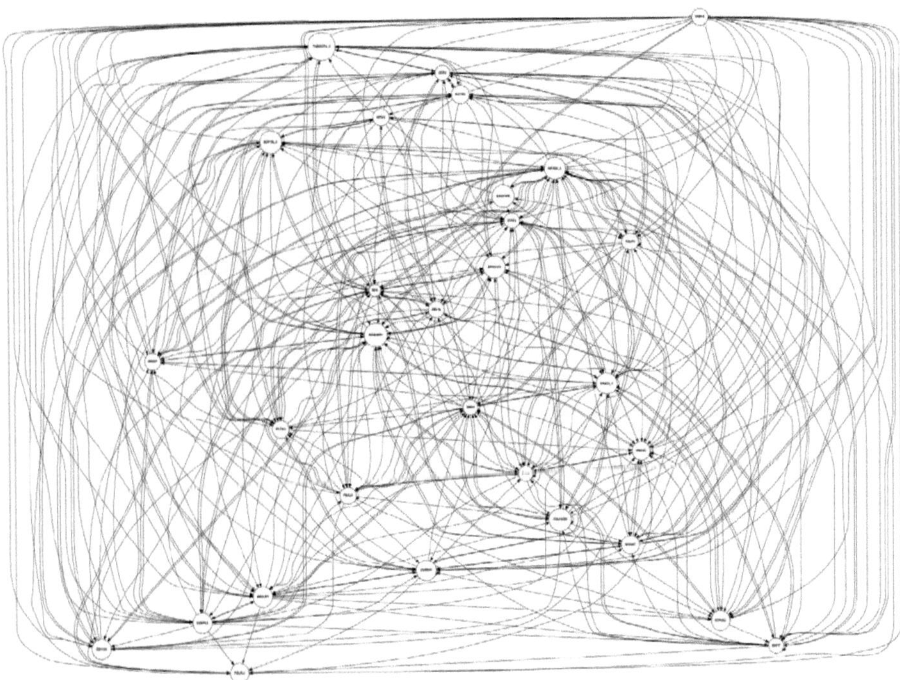

Figure 7 *USP14*-repressive molecular network of high human left hemisphere by GRNInfer. n=14. Black arrow represents the activation relationship and empty circle as the inhibition one. Con, chimpanzee left hemisphere; Ex, human left hemisphere; Inh, inhibition.

Ex-Inh-Terms and Numbers

nucleus	14	endocrine and CNS	1	insulin signaling	1	congenital hearing loss	1
protein binding	12	negative regulation of metabolism	1	cellular morphogenesis	1	usher syndrome type iii	1
cytoplasm	10	apoptosis	1	protein targeting	1	visual loss	1
RNA splicing	8	IL-6	1	Wnt Signaling	1	hearing loss sensorineural	1
cytosol	7	frontotemporal dementia with parkinsonism-17	1	EGFR1	1	vestibular diseases	1
deafness	7	attention deficit disorder | conduct disorder | oppositional defiant disorder | dementia	1	IL-4	1	usher syndrome type ig	1
membrane	6	dementia | ALS/amyotrophic lateral sclerosis	1	Wnt Signaling and pluripotency	1	usher syndrome type 1	1
zinc ion-binding	6	progressive supranuclear palsy	1	malignant neoplasms	1	hearing loss	1
nucleotide-binding	6	frontotemporal dementia and pick-like 3R and 4R tauopathy	1	neoplasms	1	sensorineural nonsyndromic	1
plasma membrane	5	inherited dementia FTDP-17	1	nephropathy	1	steroid hormone receptor activity	1
ATP-binding	5	dementia | neuropathy	1	diabetic	1	circadian exercise	1
transferase activity	5	Jakob-Creutzfeldt disease	1	binding	1	nuclear receptors	1
metal ion-binding	5	familial atypical progressive supranuclear palsy	1	Notch signaling pathway	1	DNA-directed RNA polymerase II	1
nuclear mRNA splicing	5	alzheimer's disease	1	Wnt receptor signaling pathway	1	core complex	1
via spliceosome	5	parkinson's disease	1	interspecies interaction between organisms	1	DNA-binding	1
neurosensory	5	huntington's disease | progressive supranuclear palsy | dementia	1	positive regulation of ubiquitin ligase activity during mitotic cell cycle	1	DNA-directed RNA polymerase activity	1
structural constituent of cytoskeleton	4	frontotemporal lobar degeneration	1	ubiquitin mediated proteolysis	1	protein dimerization activity	1
MAPK signaling pathway	4	posterior cortical atrophy	1	hedgehog signaling pathway	1	purine metabolism	1
regulation of transcription DNA-dependent	4	pick's disease	1	TGF-beta-receptor	1	pyrimidine metabolism	1
nucleoplasm	4	primary progressive aphasia	1	TNF-alpha-NF-kB	1	RNA polymerase	1
microtubule	3	frontotemporal dementia	1	presenilin action in notch and Wnt signaling	1	transcription from RNA polymerase II promoter	1
alzheimer's disease	3	intracellular	1	molecular function	1	nucleotidyltransferase activity	1
microtubule cytoskeleton	3	membrane fraction	1	lipid transport	1	heart failure	1
frontotemporal	3	apparatus	1	riboflavin kinase activity	1	coronary disease	1
endoplasmic reticulum	3	guanyl-nucleotide exchange factor activity	1	riboflavin biosynthesis	1	heart failure congestive	1
endoplasmic reticulum membrane	3	Ras protein signal transduction	1	riboflavin metabolism	1	kidney diseases	1
RNA-binding	3	cell differentiation	1	inositol or phosphatidylinositol kinase activity	1	cardiomegaly	1
integral to membrane	3	regulation of small GTPase mediated signal transduction	1	protein serine/threonine kinase activity	1	nucleoplasm DNA-dependent protein kinase complex	1
DNA repair	3	T cell receptor signaling pathway	1	manganese ion-binding	1	DNA-binding DNA-dependent protein kinase activity	1
response to DNA damage stimulus	3	asthma	1	mRNA catabolism	1	double-strand break repair via nonhomologous end joining	1
protein complex assembly	3	phosphatidylinositol-4,5-bisphosphate-binding	1	nonsense-mediated decay	1	DNA recombination	1
cell cycle	3	G-protein coupled receptor protein signaling pathway	1	mRNA export from nucleus	1	non-homologous end-joining	1
transcription	3	nuclear pore	1	protein amino acid autophosphorylation	1	response to endogenous stimulus	1
cytoskeleton	2	nuclear membrane	1	phosphoinositide phosphorylation	1	fas pathway and stress induction of HSP regulation Biocarta	1
microtubule associated complex	2	ATP-dependent helicase activity	1	kinase activity	1	non-homologous end joining	1
negative regulation of microtubule depolymerization	2	hydrolase activity	1	spliceosome complex	1	FAS signaling pathway (CD95)	1
microtubule based process	2	protein transport	1	nuclear speck	1	HIV-1 Nef negative effector of Fas and TNF	1
dementia	2	mRNA transport	1	mRNA splice site selection	1	TNFR1 signaling pathway	1
calcium ion-binding	2	intracellular protein transmembrane transport	1	actin-binding	1	cell cycle: G2/M checkpoint	1
diacylglycerol binding	2	folate biosynthesis	1	structural molecule activity	1	brain cancer	1
lipid-binding	2	platelet dense tubular network membrane	1	microtubule bundle formation	1	breast cancer	1
helicase activity	2	ion channel activity	1	dendrite development	1	ubiquitin ligase complex	1
signal transduction	2	inositol 1,4,5-triphosphate-sensitive calcium-release channel activity	1	muscle development	1	ubiquitin-protein ligase activity	1
G Protein Signaling	2	calcium channel activity	1	response to stress	1	protein ubiquitination	1
protein amino acid phosphorylation	2	inositol-1,4,5-triphosphate receptor activity	1	cell cycle arrest	1	modification-dependent protein catabolism	1
kinase activity	2	calcium ion transporter activity	1	negative regulation of cell growth	1	cyclin-dependent protein kinase activity	1
ubiquitin-dependent protein catabolism	2	response to hypoxia	1	regulation of cell proliferation	1	regulation of cell cycle	1
Wnt signaling pathway	2	ion transport	1	transcription factor TFIIA complex	1	spindle pole	1
ubiquitin cycle	2	calcium ion transport	1	transcription coactivator activity	1	centrosome	1
Wnt	2	cell death	1	transcription factor-binding	1	gamma-tubulin ring complex	1
nucleolus	2	calcium signaling pathway	1	general RNA polymerase II transcription factor activity	1	microtubule nucleation	1
steroid metabolism	2	phosphatidylinositol signaling system	1	TATA-binding protein-binding	1	MAPT (microtubule-associated protein tau)	
magnesium ion-binding	2	gap junction	1	protein homodimerization activity	1	RASGRP1 (RAS guanyl releasing protein 1 (calcium and DAG-regulated))	
peptidyl-serine phosphorylation	2	long-term potentiation	1	protein heterodimerization activity	1	TULP3 (tubby like protein 3)	
via transesterification reactions with bulged adenosine as nucleophile	2	long-term depression	1	positive regulation of transcription factor activity	1	DDX19A (DEAD (Asp-Glu-Ala-As) box polypeptide 19A)	
via transesterification reactions	2	GnRH signaling pathway	1	transcriptional preinitiation complex formation	1	ITPR1 (inositol 1,4,5-triphosphate receptor 1)	
mRNA processing Reactome	2	calcium regulation in cardiac cells	1	positive regulation of specific transcription from RNA polymerase II promoter	1	KIAA1109 (KIAA1109)	
mRNA processing	2	smooth muscle contraction	1	general transcription from RNA polymerase II promoter	1	PRKCI_1 (protein kinase C iota)	
negative regulation of cell proliferation	2	B cell receptor	1	basal transcription factors	1	SEL1L (sel-1 suppressor of lin-12-like (C. elegans))	
transcription initiation from RNA polymerase II promoter	2	T-cell-receptor	1	RNA polymerase II transcription factor activity	1	BTRC (beta-transducin repeat containing)	
RNA elongation from RNA polymerase II promoter	2	cadmium induces DNA synthesis and proliferation in macrophages	1	autosomal recessive 3	1	CDR2 (cerebellar degeneration-related protein 2 62kDa)	
RNA transcription Reactome	2	polarisome	1	autosomal recessive 18	1	OSBPL8 (oxysterol binding protein-like 8)	
usher syndrome	2	endosome	1	retinal degeneration	1	RFK (riboflavin kinase)	
transcription factor activity	2	apical part of cell	1	retinal diseases	1	SMG1 (SMG1 homolog phosphatidylinositol 3-kinase-related kinase (C. elegans))	
sequence-specific DNA-binding	2	protein kinase C activity	1	retinitis pigmentosa	1	SFRS1 (splicing factor arginine/serine-rich 1)	
protein modification	2	phospholipid-binding	1	autosomal recessive 9	1	MAP1B_2 (microtubule-associated protein 1B)	
axon	1	protein targeting to membrane	1	usher syndrome type ic	1	CGRRF1 (cell growth regulator with ring finger domain 1)	
growth cone	1	cytoskeleton organization and biogenesis	1	autosomal recessive 2	1	GTF2A2 (general transcription factor IIA 2 12kDa)	
cell projection	1	actin filament organization	1	autosomal dominant nonsyndromic sensorineural 11	1	MEIS3P1 (Meis homeobox 3 pseudogene 1)	
tubulin	1	intracellular signaling cascade	1	hearing disorders	1	NR1D2_2 (nuclear receptor subfamily 1 group D member 2)	
microtubule-binding	1	membrane organization and biogenesis	1	congenital deafness	1	POLR2J (polymerase (RNA) II (DNA directed) polypeptide J 13.3kDa)	
lipoprotein-binding	1	vesicle-mediated transport	1	usher syndrome type id	1	PRKDC (protein kinase DNA-activated catalytic polypeptide)	
SH3 domain-binding	1	establishment of apical/basal cell polarity	1	hereditary diseases	1	SFRS7 (splicing factor arginine/serine-rich 7 35kDa)	
enzyme-binding	1	eye photoreceptor cell development	1	autosomal recessive 1	1	FBXL5 (F-box and leucine-rich repeat protein 5)	
identical protein-binding	1	establishment and/or maintenance of epithelial cell polarity	1	abnormal vision	1	PCTK2 (PCTAIRE protein kinase 2)	
apolipoprotein E-binding	1	intercellular junction assembly and maintenance	1	blindness	1	TUBGCP4_2 (tubulin gamma complex associated protein 4)	
microtubule cytoskeleton organization and biogenesis	1	secretion	1	autosomal recessive 4	1	AF052119 (solute carrier family 25 (mitochondrial carrier; adenine nucleotide translocator) member 4)	
positive regulation of microtubule polymerization	1	vesicle budding	1	usher syndrome type if	1	KIAA0888 (family with sequence similarity 169 member A)	
positive regulation of axon extension	1	tight junction	1	usher syndrome type iia	1	M19267 (Human tropomyosin mRNA complete cds)	
generation of neurons	1	insulin signaling pathway	1	deafness sensorineural	1	R37702 (yf70d02.s1 Soares infant brain 1NIB Homo sapiens cDNA clone IMAGE:25627 3- mRNA sequence)	

Figure 8 *USP14*-repressive terms and occurrence numbers in high human left hemisphere based on GO, KEGG, GenMAPP, BioCarta and Disease via MAS 3.0. Ex, human left hemisphere; Inh, inhibition.

Discussion

USP14-repressive different molecular Pearson mutual-positive-correlation network was setup in high human left hemisphere compared with the corresponding low chimpanzee left hemisphere (Fig. 7).

Our identified *USP14* feedback-inhibitive molecular network showed *MAPT* (microtubule-associated protein tau), *RASGRP1* (RAS guanyl releasing protein 1 (calcium and DAG-regulated)), *RFK* (riboflavin kinase), *AF052119* (solute carrier family 25 (mitochondrial carrier; adenine nucleotide translocator) member 4), *M19267* (Human tropomyosin mRNA complete cds) in high human left hemisphere. *USP14* feedback-inhibitive terms network includes cytoplasm, cytosol, plasma membrane, cell projection, membrane, intracellular, membrane fraction, endoplasmic reticulum, endoplasmic reticulum membrane; axon, positive regulation of axon extension, generation of neurons, endocrine and CNS, dementia | neuropathy, T cell receptor signaling pathway; microtubule, microtubule associated complex, microtubule-binding, microtubule cytoskeleton organization and biogenesis, negative regulation of microtubule depolymerization, positive regulation of microtubule polymerization, microtubule cytoskeleton, apoptosis, microtubule-based process, cell differentiation; lipoprotein-binding, SH3 domain-binding, enzyme-binding, identical protein-binding, apolipoprotein E-binding, guanyl-nucleotide exchange factor activity, calcium ion-binding, protein-binding, zinc ion-binding, diacylglycerol-binding, nucleotide-binding, magnesium ion-binding, ATP-binding, riboflavin kinase activity, transferase activity in high human left hemisphere based on integrative GO, KEGG, GenMAPP, BioCarta and disease databases (Fig. 8). Therefore, we propose high *USP14* feedback-inhibitive axon, T cell induced microtubule complex through *MAPT-RASGRP1-RFK-AF052119-M19267* in human left hemisphere.

Microtubule positive relationship with solute carrier, tropomyosin has been reported in references as follows: Solute carrier family 3 member 2 (Slc3a2) controls yolk syncytial layer (YSL) formation by regulating microtubule networks in the zebrafish embryo [31]. Tropomyosin reverses cross-linking of F-actin with microtubule-associated protein-2 [32].

In summary, *USP14*-repressive molecular Pearson mutual-positive-correlation network was constructed in high human left hemisphere from the overlapping molecules of GRNInfer with Pearson. We propose and verify high *USP14* feedback-inhibitive axon, T cell induced microtubule complex through *MAPT-RASGRP1-RFK-AF052119-M19267* in human left hemisphere. New critical insights and perspectives for the future will be further verified our hypothesis by experimental biology. High *USP14* feedback-inhibitive axon, T cell induced microtubule complex through *MAPT-RASGRP1-RFK-AF052119-M19267* is very useful to identify novel markers and potential drugs for prognosis and therapy, develop a new route for studying the pathogenesis.

References

[1] Flach, K., E. Ramminger, I. Hilbrich, et al. Axotrophin/MARCH7 acts as an E3 ubiquitin ligase and ubiquitinates tau protein in vitro impairing microtubule binding. Biochim Biophys Acta. [J] 1842(9). 2014. 1527-38.

[2] Johnson, J.L., C. Lu, E. Raharjo, et al. Levels of the ubiquitin ligase substrate adaptor MEL-26 are inversely correlated with MEI-1/katanin microtubule-severing activity during both meiosis and mitosis. Dev Biol. [J] 330(2). 2009. 349-57.

[3] Lu, C. and P.E. Mains. The C. elegans anaphase promoting complex and MBK-2/DYRK kinase act

redundantly with CUL-3/MEL-26 ubiquitin ligase to degrade MEI-1 microtubule-severing activity after meiosis. Dev Biol. [J] 302(2). 2007. 438-47.

[4] Chuang, M., A. Goncharov, S. Wang, et al. The Microtubule Minus-End-Binding Protein Patronin/PTRN-1 Is Required for Axon Regeneration in C. elegans. Cell Rep. [J] 9(3). 2014. 874-83.

[5] Liu, G. and T. Dwyer. Microtubule dynamics in axon guidance. Neurosci Bull. [J] 30(4). 2014. 569-83.

[6] Liz, M.A., F.M. Mar, T.E. Santos, et al. Neuronal deletion of GSK3beta increases microtubule speed in the growth cone and enhances axon regeneration via CRMP-2 and independently of MAP1B and CLASP2. BMC Biol. [J] 12. 2014. 47.

[7] Nwagbara, B.U., A.E. Faris, E.A. Bearce, et al. TACC3 is a microtubule plus end-tracking protein that promotes axon elongation and also regulates microtubule plus end dynamics in multiple embryonic cell types. Mol Biol Cell. [J] 25(21). 2014. 3350-62.

[8] Baratt, A., S.N. Arkhipov, and I.V. Maly. An experimental and computational study of effects of microtubule stabilization on T-cell polarity. PLoS One. [J] 3(12). 2008. e3861.

[9] Dong, B., S.S. Zhang, W. Gao, et al. Mammalian diaphanous-related formin 1 regulates GSK3beta-dependent microtubule dynamics required for T cell migratory polarization. PLoS One. [J] 8(11). 2013. e80500.

[10] Lin, J., K.K. Hou, H. Piwnica-Worms, et al. The polarity protein Par1b/EMK/MARK2 regulates T cell receptor-induced microtubule-organizing center polarization. J Immunol. [J] 183(2). 2009. 1215-21.

[11] Storey., J.D. A direct approach to false discovery rates. J. Roy. Stat. Soc., Ser. B. [J] 64. 2002. 479–498.

[12] Wang, Y., T. Joshi, X.S. Zhang, et al. Inferring gene regulatory networks from multiple microarray datasets. Bioinformatics. [J] 22(19). 2006. 2413-20.

[13] Wang, L., J. Huang, M. Jiang, et al. Activated PTHLH Coupling Feedback Phosphoinositide to G-Protein Receptor Signal-Induced Cell Adhesion Network in Human Hepatocellular Carcinoma by Systems-Theoretic Analysis. ScientificWorldJournal. [J] 2012. 2012. 428979.

[14] Wang, L., J. Huang, M. Jiang, et al. Inhibited PTHLH downstream leukocyte adhesion-mediated protein amino acid N-linked glycosylation coupling Notch and JAK-STAT cascade to iron-sulfur cluster assembly-induced aging network in no-tumor hepatitis/cirrhotic tissues (HBV or HCV infection) by systems-theoretical analysis. Integr Biol (Camb). [J] 4(10). 2012. 1256-62.

[15] Wang, L., J. Huang, M. Jiang, et al. Tissue-specific transplantation antigen P35B (TSTA3) immune response-mediated metabolism coupling cell cycle to postreplication repair network in no-tumor hepatitis/cirrhotic tissues (HBV or HCV infection) by biocomputation. Immunol Res. [J] 52(3). 2012. 258-68.

[16] Wang, L., J. Huang, M. Jiang, et al. Signal transducer and activator of transcription 2 (STAT2) metabolism coupling postmitotic outgrowth to visual and sound perception network in human left cerebrum by biocomputation. J Mol Neurosci. [J] 47(3). 2012. 649-58.

[17] Lin, H., L. Wang, M. Jiang, et al. P-glycoprotein (ABCB1) inhibited network of mitochondrion transport along microtubule and BMP signal-induced cell shape in chimpanzee left cerebrum by systems-theoretical analysis. Cell Biochem Funct. [J] 30(7). 2012. 582-7.

[18] Huang, J., L. Wang, M. Jiang, et al. PTHLH coupling upstream negative regulation of fatty acid biosynthesis and Wnt receptor signal to downstream peptidase activity-induced apoptosis network in human hepatocellular carcinoma by systems-theoretical analysis. J Recept Signal Transduct Res. [J] 32(5). 2012. 250-6.

[19] Wang, L., L. Sun, J. Huang, et al. Cyclin-dependent kinase inhibitor 3 (CDKN3) novel cell cycle computational network between human non-malignancy associated hepatitis/cirrhosis and hepatocellular carcinoma (HCC) transformation. Cell Prolif. [J] 44(3). 2011. 291-9.

[20] Wang, L., J. Huang, M. Jiang, et al. AFP computational secreted network construction and analysis between human hepatocellular carcinoma (HCC) and no-tumor hepatitis/cirrhotic liver tissues. Tumour Biol. [J] 31(5). 2011. 417-25.

[21] Wang, L., J. Huang, M. Jiang, et al. Survivin (BIRC5) cell cycle computational network in human no-tumor hepatitis/cirrhosis and hepatocellular carcinoma transformation. J Cell Biochem. [J] 112(5). 2011. 1286-94.

[22] Wang, L., J. Huang, M. Jiang, et al. MYBPC1 computational phosphoprotein network construction and analysis between frontal cortex of HIV encephalitis (HIVE) and HIVE-control patients. Cell Mol Neurobiol. [J] 31(2). 2011. 233-41.

[23] Wang, L., J. Huang, and M. Jiang. CREB5 computational regulation network construction and analysis between frontal cortex of HIV encephalitis (HIVE) and HIVE-control patients. Cell Biochem Biophys. [J] 60(3). 2011. 199-207.

[24] Wang, L., J. Huang, and M. Jiang. RRM2 computational phosphoprotein network construction and analysis between no-tumor hepatitis/cirrhotic liver tissues and human hepatocellular carcinoma (HCC). Cell Physiol Biochem. [J] 26(3). 2011. 303-10.

[25] Sun, L., L. Wang, M. Jiang, et al. Glycogen debranching enzyme 6 (*AGL*), enolase 1 (ENOSF1), ectonucleotide pyrophosphatase 2 (ENPP2_1), glutathione S-transferase 3 (*GSTM3*_3) and mannosidase (MAN2B2) metabolism computational network analysis between chimpanzee and human left cerebrum. Cell Biochem Biophys. [J] 61(3). 2011. 493-505.

[26] Huang, J.X., L. Wang, and M.H. Jiang. TNFRSF11B computational development network construction and analysis between frontal cortex of HIV encephalitis (HIVE) and HIVE-control patients. J Inflamm (Lond). [J] 7. 2011. 50.

[27] Sun, Y., L. Wang, M. Jiang, et al. Secreted Phosphoprotein 1 Upstream Invasive Network Construction and Analysis of Lung Adenocarcinoma Compared with Human Normal Adjacent

Tissues by Integrative Biocomputation. Cell Biochem Biophys. [J] 56(2-3). 2010. 59-71.

[28] Huang, J., L. Wang, M. Jiang, et al. Interferon α-Inducible Protein 27 Computational Network Construction and Comparison between the Frontal Cortex of HIV Encephalitis (HIVE) and HIVE-Control Patients The Open Genomics Journal [J] 3(1875-693X). 2010. 1-8.

[29] Wang, L., Y. Sun, M. Jiang, et al. Integrative decomposition procedure and Kappa statistics for the distinguished single molecular network construction and analysis. J Biomed Biotechnol. [J] 2009. 2009. 726728.

[30] Wang, L., Y. Sun, M. Jiang, et al. FOS proliferating network construction in early colorectal cancer (CRC) based on integrative significant function cluster and inferring analysis. Cancer Invest. [J] 27(8). 2009. 816-24.

[31] Takesono, A., J. Moger, S. Farooq, et al. Solute carrier family 3 member 2 (Slc3a2) controls yolk syncytial layer (YSL) formation by regulating microtubule networks in the zebrafish embryo. Proc Natl Acad Sci U S A. [J] 109(9). 2012. 3371-6.

[32] Okagaki, T. and S. Asakura. Tropomyosin reverses cross-linking of F-actin with microtubule-associated protein-2. J Biochem. [J] 101(1). 1987. 189-97.

Chapter 3: High *CCRK* inside-out inhibiting DNA damage induced mitotic spindle through *CLCN4-MIA3-CDKN1B-KPNB1-PSMA4-SMG1-AGL*

Abstract

25 different Pearson mutual-positive-correlation *CCRK*-repressive molecular network was constructed from 83 overlapping of 274 GRNInfer and 144 Pearson under *CCRK* CC ≤ -0.25 in high human left hemisphere compared with low chimpanzee left hemisphere. Our identified *CCRK* inside-out inhibiting molecular network showed *CLCN4* (chloride channel 4), *MIA3* (melanoma inhibitory activity family member 3), *CDKN1B* (cyclin-dependent kinase inhibitor 1B (p27 Kip1)), *KPNB1* (karyopherin (importin) beta 1), *PSMA4* (proteasome (prosome macropain) subunit alpha 4), *SMG1* (*SMG1* homolog phosphatidylinositol 3-kinase-related kinase (C. elegans)), *AGL* (amylo-1 6-glucosidase 4-alpha-glucanotransferase) in high human left hemisphere. *CCRK* inside-out inhibiting terms network includes membrane, integral to membrane, nucleus, cytoplasm, cytosol, protein import into nucleus, NLS-bearing substrate import into nucleus, ribosomal protein import into nucleus, endomembrane system, nucleocytoplasmic transport, mechanism of protein import into the nucleus, mRNA export from nucleus, extracellular space; wound healing, autoimmune thyroid disease | thyroid disease, autoimmune, autoimmune diseases, response to DNA damage stimulus; role of ran in mitotic spindle regulation, positive regulation of leukocyte migration, negative regulation of cell adhesion, negative regulation

of cell migration, G1/S transition of mitotic cell cycle, induction of apoptosis, cell cycle, cell cycle arrest, negative regulation of cell proliferation, negative regulation of cell growth, autophagic cell death, regulation of cell proliferation, cell cycle-G1 to S control Reactome, PTEN dependent cell cycle arrest and apoptosis, CDK regulation of DNA replication, cell cycle: G1/S check point, cyclins and cell cycle regulation, regulation of p27 Phosphorylation during cell cycle progression, negative regulation of ubiquitin ligase activity during mitotic cell cycle, positive regulation of ubiquitin ligase activity during mitotic cell cycle based on integrative GO, KEGG, GenMAPP, BioCarta and disease databases in high human left hemisphere. Therefore, we propose high *CCRK* inside-out inhibiting DNA damage induced mitotic spindle through *CLCN4-MIA3-CDKN1B-KPNB1-PSMA4-SMG1-AGL* in human left hemisphere.

Keywords: *CCRK* inhibiting network; inside-out; mitotic spindle; DNA damage

Introduction

CCRK is one of our identified significant molecules (fold change 2) in high human left hemisphere compared with the corresponding low chimpanzee left hemisphere. *CCRK* appears nucleus; nucleotide-binding, cyclin-dependent protein kinase activity, ATP-binding, transferase activity; protein amino acid phosphorylation, cell cycle, cell division; protein-tyrosine kinase activity; malignant neoplasm of breast, neoplasms based on GO, KEGG, GenMAPP, BioCarta and disease databases. Cyclin-dependent kinase negative relationship with spindle has been reported in references as follows: . Novel functional assay for spindle-assembly checkpoint by cyclin-dependent kinase activity to predict taxane chemosensitivity in breast tumor patient; Intrinsic and cyclin-dependent kinase-dependent control of spindle pole body duplication in budding yeast; The spindle checkpoint requires cyclin-dependent kinase activity [1-3]. DNA damage relation with spindle has been reported in references as follows: TPX2: of spindle assembly, DNA damage response, and cancer; Mdb1, a fission yeast homolog of human MDC1, modulates DNA damage response and mitotic spindle function [4, 5]. Yet high *CCRK* inside-out inhibiting DNA damage induced mitotic spindle through *CLCN4-MIA3-CDKN1B-KPNB1-PSMA4-SMG1-AGL* in human left hemisphere is not clear.

25 different Pearson mutual-positive-correlation *CCRK*-repressive molecular network was constructed from 83 overlapping of 274 GRNInfer and

144 Pearson under *CCRK* CC ≤ -0.25 in high human left hemisphere compared with low chimpanzee left hemisphere (Fig. 9A-9D).

A

Ex-Inh-GRNInfer
AA975427, ABCB1, ABCC10, ACOX3, ACTR2, AF052141, AGL, AL049242, AL080232, ALDH7A1, ALG8, ARHGAP12, ATP2B1, ATP5J, B3GNT1, BAG5, BCAS1, BEST1, BLCAP, BMP2K, BRP44, BTN3A3, BTRC, C16orf35, C1orf61_2, C1orf63, CA2, CACNB3, CALM1, CAMTA1, CBLB, CCNO, 2008_9_8, AF052119, AL031282, AL049987, ANAPC10, ATP5J2, BAP1, C10orf10, C14orf1, C18orf10, C1orf61_1, CD59, CDH19, CDR2, CFHR1, CFLAR, CGRRF1, CHAD, CLDN10, CLTB, COL6A1, COL6A2, CTB_1048E9.5, CTNNA1_1, CTNNA1_2, CYP2J2, DDX3Y, DGCR5, DKFZp434H1419, DLEU1, DZIP3, EXTL2, FABP6, FBXL5, FGF1, GAB2, GABRB1, GEM, GM2A, GOLGA3, GPR68, GSTM3_3, GSTM5, H10776, HDDC2, HERC2P2, HLA_DPB1, HNRNPA0, HNRPDL, HOMER1, HSD17B6, HSP90AB1, HSPA1A, HYPE, IFI44L, INHBB, KCNK1, KIAA0888, KIAA0895, L12535, LAMP2, LGALS3BP, LIMCH1, LOC157627, LOH11CR2A, LRRC37A, LRRC42, M19267, MAGOH, MAN2B2, MAP1B_1, MAPT, MED6, MGC15523, MOAP1, MTMR1, NR1D2_1, NR1D2_2, NSUN5C, NUPR1, OGFR, PAK3, PAX6, PCDHGA8, PCTK2, PDE4DIP, PDE8A, PDIA2, PDLIM5, PDPN, PIN1, PMS2L1, PPARD, PRDX6, PRKCI_1, PTTG1IP, R37702, RAD50, RAD51C, RAPGEF4, RASA4, RASGRP1, RBM34, RCHY1, RFK, RFPL1S, RNF2, ROM1, RPA3, SAPS2, SC4MOL, SELENBP1, SFPQ_1, SH3BGR, SLC13A3, SLC25A6, SMAD1_1, SMAD1_2, SNRPE, SRI, SSH1, SSTR2, STAT2, TCF25, TFCP2, TH, THBD, TIGR:, TSPYL2, U59632, UBR5, UNG, USP22, VDAC2, W22289, W28807, WIPF2, X82895, ZBTB43, ZNF174, ZNF44, ZNF443, CDKN1B, CLCN4, DHCR24, DIXDC1, DTNA, EID1, EIF1AY, ENOSF1, ENPP2_2, EPM2AIP1, EVI5, FAM131B, FCMD, FDFT1, FEZ1, FLJ43806, FLNB, FOXN3_1, FOXN3_2, GAPVD1, GIPC3, GLO1, GOSR1, GPD1, GSTM3_2, GTF2A2, GTF2I_1, GTF2I_2, GUSBP1, H24861, HCN2, HERC2P3, HNRNPA1, HNRPH3, HSPA2, IMPA1, IPO13, IQCK, ITGB3BP, JARID1D, KIAA0644, KPNB1, KRT10, LRPPRC, MAF, MARCKSL1, MIA3, MTUS1, NAIP, NPC1, OGG1_1, OGG1_2, OPTN, OSBPL8, P4HB, PALLD, PHKG1, PHLDB1, PHYH, POLR2J, POLR2J3, PPP1R13B, PRKCI_2, PSMA4, RAB7L1, RASSF2, RBBP6, RCBTB2, RECQL, RPL23AP7, SAMHD1, SCYE1, SF3B3, SFRS7, SLC25A16, SLC25A46, SMG1, SPTLC1, SQSTM1, SULT1A2, TAF1C, THBS2, TJP2, TMED10, TOR1B, TXNL4A, TYRO3, U00928, UBB, USP46, W27641, WASF3, WDFY3, WWOX, X81789, Z75311, ZNF254, ZNF271, ZNF294, ZNF91_2

B

Ex-Inh-Pearson
ELNR1, AA975427, AB002448, ACADM, AGL, AK3P1, AL042668, AL049242, AL049278, AL080234, ARF6, ARHGAP12, AW043812, BRP44, C16orf35, CAMTA1, CBLB, CBR4, CDC25B, CDKN1B, CDR2, CFLAR, CGRRF1, CHAD, CLASP2, CLCN4, COL6A1, CTBP1, CTRL, DDX19A, DICER1, DIXDC1, EIF1AY, EPHX2, EPM2AIP1, FAM114A1, FBXL5, FCMD, FKBPL, FLJ43806, FLNB, FOXN3_1, FOXN3_2, GAPVD1, GATAD1, GDF10, GIPC3, GPD1, GTF2A2, GUSBP1, H2AFX, HCN2, HDDC2, HERC2P3, HNRNPA0, HNRNPA1, HNRPDL, HNRPH3, HSD17B6, INSIG1, ITPR1, KCNK1, KIAA0368, KIAA0895, KIAA1109, KPNB1, KRT10, LGALS3, MED13, MED6, MEIS3P1, MIA3, MTHFS, MTUS1, N58318, NAIP, NDRG4, NFIB, NPAL3, NR1D2_1, NR1D2_2, NUPR1, OGG1_1, OGG1_2, OSBPL8, PAX6, PCBP2_1, PCBP2_2, PCDHGA8, PDPN, PEPD, PHKG1, POLR2J, PRKCI_1, PSMA4, R37702, RAB7L1, RASGRP1, RBCK1, RBL2, RCHY1, RECQL, RNF14, RNF2, RP4_691N24.1, RPS26, SARM1, SCYE1, SEL1L, SF3B3, SFPQ_1, SFPQ_2, SFRS1, SFRS7, SLC39A6, SMAD1_1, SMAD1_2, SMC5, SMG1, SNW1, SPAG9, STAT2, TERF1_1, TERF1_2, THBD, TJP2, TLOC1, TMEM41B, TOR1B, TULP3, U00928, U59632, UBXD2, UPF3A, VBP1, VDAC2, W27641, WDFY3, WDR57, X81789, ZNF271, ZNF451, ZNF91_1, ZNF91_2

C

Ex-Inh-Overlap-GRNInfer and Pearson
AA975427, AGL, AL049242, ARHGAP12, BRP44, C16orf35, CAMTA1, CBLB, CDKN1B, CDR2, CFLAR, CGRRF1, CHAD, CLCN4, COL6A1, DIXDC1, EIF1AY, EPM2AIP1, FBXL5, FCMD, FLJ43806, FLNB, FOXN3_1, FOXN3_2, GAPVD1, GIPC3, GPD1, GTF2A2, GUSBP1, HCN2, HDDC2, HERC2P3, HNRNPA0, HNRNPA1, HNRPDL, HNRPH3, HSD17B6, KCNK1, KIAA0895, KPNB1, KRT10, MED6, MIA3, MTUS1, NAIP, NR1D2_1, NR1D2_2, NUPR1, OGG1_1, OGG1_2, OSBPL8, PAX6, PCDHGA8, PDPN, PHKG1, POLR2J, PRKCI_1, PSMA4, R37702, RAB7L1, RASGRP1, RCHY1, RECQL, RNF2, SCYE1, SF3B3, SFPQ_1, SFRS7, SMAD1_1, SMAD1_2, SMG1, STAT2, THBD, TJP2, TOR1B, U00928, U59632, VDAC2, W27641, WDFY3, X81789, ZNF271, ZNF91_2

D

Ex-Inh-Different Mutual Positive Pearson Correlation Compared with Con
RASGRP1, CLCN4, MIA3, KCNK1, CAMTA1, CDKN1B, KPNB1, PSMA4, RCHY1, SMG1, EPM2AIP1, TOR1B, AGL, FOXN3_2, GTF2A2, HNRNPA0, NR1D2_1, NR1D2_2, SF3B3, SFRS7, ZNF91_2, GIPC3, GUSBP1, KIAA0895, R37702

Figure 9 (A) *CCRK*-repressive molecules of high human left hemisphere by GRNInfer. **(B)** *CCRK*-repressive molecules of high human left hemisphere by Pearson. **(C)** *CCRK*-repressive overlapping molecules of high human left hemisphere by GRNInfer and Pearson. **(D)** *CCRK*-repressive different mutual-positive-correlation molecules in high human left hemisphere compared with the corresponding low chimpanzee left hemisphere. Con, chimpanzee left hemisphere; Ex, human left hemisphere; Inh, inhibition.

Materials and Methods

441 significant molecules were identified from 12558 genes of 14 high human compared with 15 low chimpanzee left hemisphere in GEO data set GDS2678 (http://www.ncbi.nlm.nih.gov/sites/GDSbrowser?acc=GDS2678) containing brain cerebrum, anterior cingulated, anterior inferior parietal, anterior inferior temporal, middle frontal gyrus, frontal pole, etc. for studying high *CCRK* inside-out inhibiting DNA damage induced mitotic spindle through *CLCN4-MIA3-CDKN1B-KPNB1-PSMA4-SMG1-AGL* using significant analysis of microarrays (SAM) (http://www-stat.stanford.edu/~tibs/SAM/) [6]. The raw microarray data were processed by log base 2. Two classes were unpaired and minimum fold change ⩾2 selected (the false-discovery rate 0%).

Gene expression values of *CCRK-repressive* different molecules were computed in high human left hemisphere compared with the corresponding low chimpanzee left hemisphere by AVERAGE and STDEV.

CCRK-repressive different mutual-positive-correlation molecular Pearson coefficients were computed in high human left hemisphere compared with the corresponding low no-tumor hepatitis/cirrhotic tissues under *CCRK* CC ≤ -0.25, as measurements of the correlation (linear dependence) including two variables X and Y and giving an inclusive value between −1 and +1.

CCRK-repressive molecular network was further constructed in high human left hemisphere by GRNInfer [7], GVedit tool and our articles [8-25].

GRNInfer is a novel mathematic method called GNR (Gene Network Reconstruction tool) based on linear programming and a decomposition procedure for inferring gene networks. The method theoretically ensures the derivation of the most consistent network structure with respect to all of the datasets, thereby not only significantly alleviating the problem of data scarcity but also remarkably improving the reconstruction reliability. The following Equation (1) represents all of the possible networks for the same dataset. We established network based on the top 441 distinguished genes and selected parameters as lambda 0.0, threshold 1.0e-10.

$$J = (X' - A)U\Lambda^{-1}V^T + YV^T = \hat{J} + YV^T \qquad (1)$$

CCRK-repressive molecular knowledge network was further calculated in high human left hemisphere based on terms and occurrence numbers of GO (Cellular Component, Molecular Function and Biological Process), KEGG, GenMAPP, BioCarta and Disease via Molecule Annotation System, MAS (CapitalBio Corporation, Beijing, China; http://bioinfo.capitalbio.com/mas3/). The primary databases of MAS integrated various well-known biological resources, such as Gene Ontology (http://www.geneontology.org), KEGG (http://www.genome.jp/kegg/), BioCarta (http://www.biocarta.com/), GenMapp (http://www.genmapp.org/), HPRD (http://www.hprd.org/), etc.

Result

CCRK-repressive different Pearson mutual-positive-correlation molecular gene expression values were illustrated column diagrams by AVERAGE and STDEV in high human left hemisphere and the corresponding low chimpanzee left hemisphere, as shown in Fig. 10A.

CCRK-repressive different mutual-positive-correlation molecular Pearson coefficients were illustrated column diagrams in high human left hemisphere compared with the corresponding low chimpanzee left hemisphere, as shown in Fig. 10B and 10C, respectively.

CCRK-repressive molecular network was further constructed by GRNInfer in high human left hemisphere, as shown in Fig. 11.

CCRK-repressive knowledge terms network was further identified by MAS 3.0 in high human left hemisphere, as shown in Fig. 12.

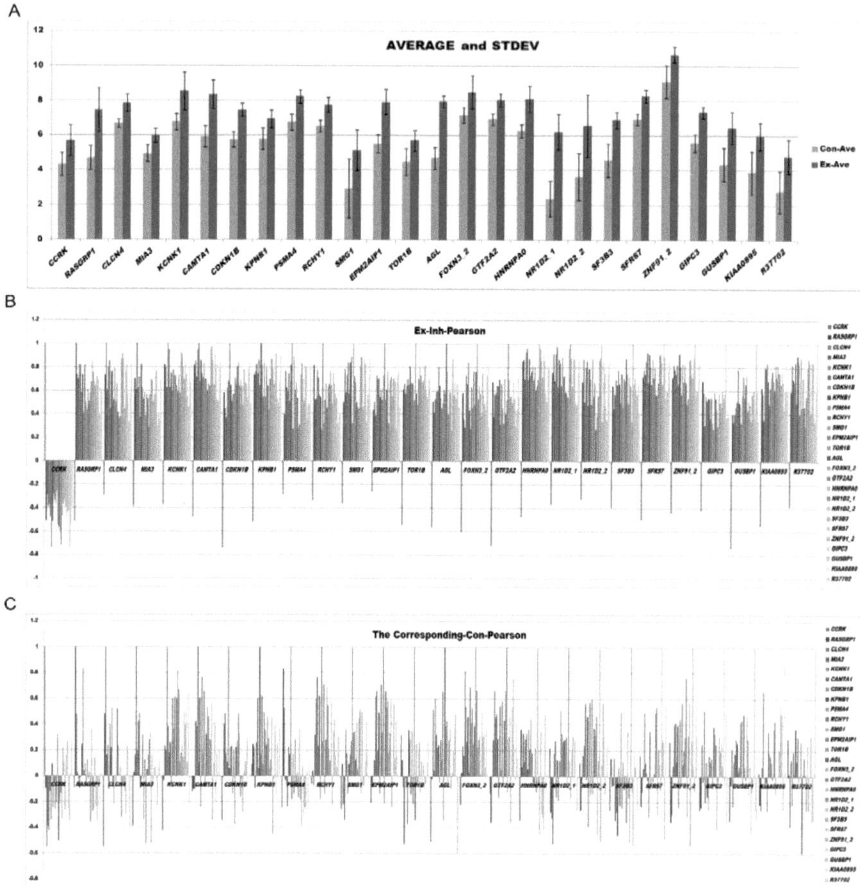

Figure 10 (A) Gene expression values of *CCRK*-repressive different mutual Pearson positive correlation molecules in high human left hemisphere and the corresponding low chimpanzee left hemisphere by AVERAGE and STDEV measurement. **(B)** Vertical quantification chart of *CCRK*-repressive different molecular mutual Pearson positive correlation coefficients in high human left hemisphere. n=14. **(C)** The corresponding correlation coefficients in low chimpanzee left hemisphere. n=15. Con, chimpanzee left hemisphere; Ex, human left hemisphere; Inh, inhibition.

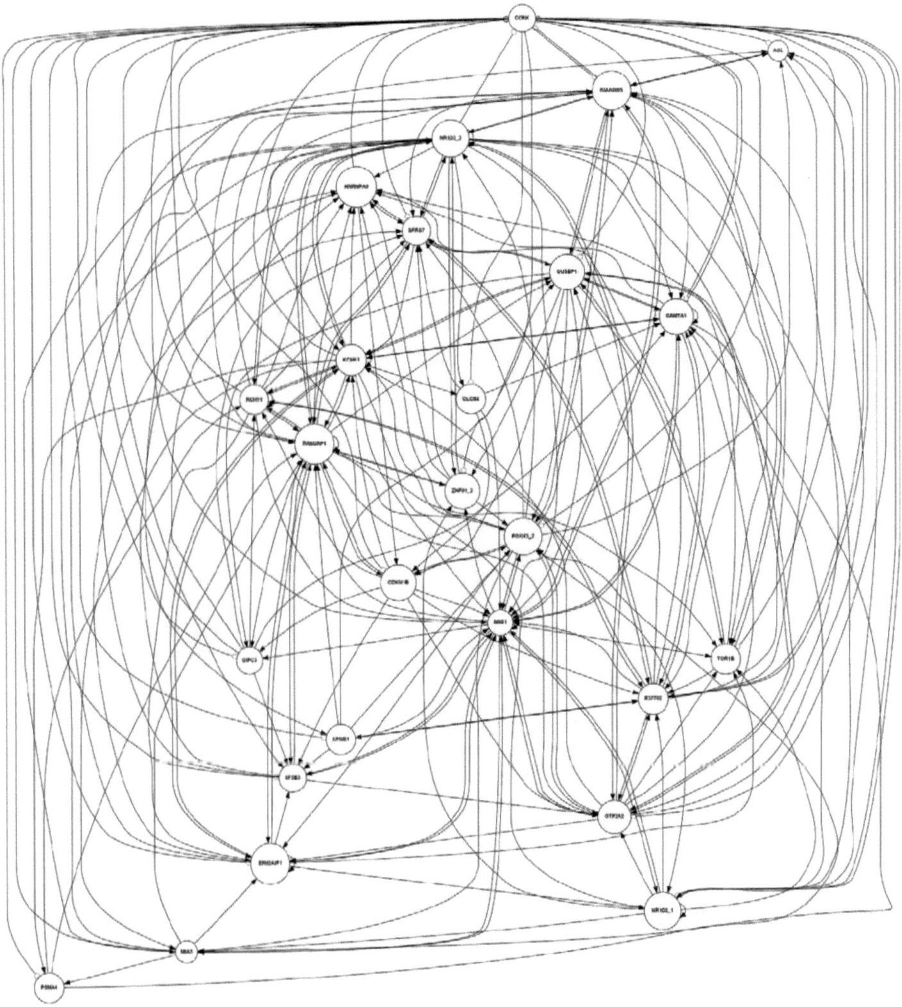

Figure 11 *CCRK*-repressive molecular network of high human left hemisphere by GRNInfer. n=14. Black arrow represents the activation relationship and empty circle as the inhibition one. Con, chimpanzee left hemisphere; Ex, human left hemisphere; Inh, inhibition.

Ex-Inh-Terms and Numbers									
nucleus	14	regulation of small GTPase mediated signal transduction	1	cholera	1	shock	1	autosomal recessive 9	1
protein-binding	8	T cell receptor signaling pathway	1	nuclear pore	1	primary carcinoma of the liver cells	1	usher syndrome type ic	1
RNA splicing	8	asthma	1	nuclear localization sequence-binding	1	p53 signaling pathway	1	autosomal recessive 2	1
cytoplasm	7	voltage-gated chloride channel activity	1	protein domain specific-binding	1	ubiquitin mediated proteolysis	1	autosomal dominant nonsyndromic sensorineural 11	1
zinc ion-binding	7	chloride ion-binding	1	docking	1	androgen-receptor	1	hearing disorders	1
deafness	7	chloride transport	1	translocation	1	inositol or phosphatidylinositol kinase activity	1	congenital deafness	1
transcription	6	anion transport	1	NLS-bearing substrate import into nucleus	1	protein serine/threonine kinase activity	1	usher syndrome type id	1
metal ion-binding	6	inorganic anion transport	1	ribosomal protein import into nucleus	1	manganese ion-binding	1	autosomal recessive 1	1
regulation of transcription DNA-dependent	5	kidney calculi	1	interspecies interaction between organisms	1	mRNA catabolism	1	abnormal vision	1
cytosol	5	dent disease	1	endomembrane system	1	nonsense-mediated decay	1	blindness	1
neurosensory	5	albinism ocular	1	protein transport	1	DNA repair	1	autosomal recessive 4	1
nuclear mRNA splicing	5	proteinuria	1	nuclear transport	1	mRNA export from nucleus	1	usher syndrome type if	1
via spliceosome	5	hemorrhagic telangiectasia	1	protein targeting	1	peptidyl-serine phosphorylation	1	usher syndrome type iia	1
membrane	4	hereditary	1	nucleocytoplasmic transport	1	protein amino acid autophosphorylation	1	deafness sensorineural	1
voltage-gated ion channel activity	4	intracranial aneurysm	1	peptide-binding	1	phosphoinositide phosphorylation	1	congenital hearing loss	1
ion transport	4	cerebral arteriopathy	1	TGF-beta-receptor	1	kinase activity	1	usher syndrome type iii	1
cell cycle	4	subarachnoid hemorrhage	1	role of ran in mitotic spindle regulation	1	endoplasmic reticulum lumen	1	visual loss	1
transcription factor activity	4	aneurysm	1	mechanism of protein import into the nucleus	1	ATP-binding	1	hearing loss sensorineural	1
endoplasmic reticulum	3	intracranial	1	graves disease	1	nucleoside-triphosphatase activity	1	vestibular diseases	1
integral to membrane	3	brain hemorrhage | brain aneurysm	1	aggressiveness	1	response to unfolded protein	1	usher syndrome type ig	1
nucleoplasm	3	hypertension	1	diabetes mellitus insulin-dependent	1	chaperone cofactor-dependent protein folding	1	usher syndrome type 1	1
circadian exercise	3	stroke	1	malignant neoplasm of pancreas	1	protein folding	1	hearing loss	1
transferase activity	3	positive regulation of leukocyte migration	1	neoplasms	1	isoamylase complex	1	sensorineural nonsyndromic	1
nucleotide-binding	3	negative regulation of cell adhesion	1	aggression	1	4-alpha-glucanotransferase activity	1	heterogeneous nuclear ribonucleoprotein complex	1
hydrolase activity	3	negative regulation of cell migration	1	breast carcinoma	1	amylo-alpha-16-glucosidase activity	1	spliceosome complex	1
carbohydrate metabolism	3	wound healing	1	diabetes type i	1	transferring glycosyl groups	1	small nuclear ribonucleoprotein complex	1
sequence-specific DNA-binding	3	hexose metabolism	1	autoimmune thyroid disease | thyroid disease	1	cation-binding	1	nucleic acid-binding	1
mRNA processing	3	voltage-gated potassium channel complex	1	autoimmune	1	glycogen biosynthesis	1	RNA splicing factor activity	1
intracellular	2	inward rectifier potassium channel activity	1	thyroid cancer	1	metabolism	1	transesterification mechanism	1
MAPK signaling pathway	2	potassium ion-binding	1	proteasome core complex (sensu Eukaryota)	1	starch and sucrose metabolism	1	hydrolyzing O-glycosyl compounds	1
hereditary hemorrhagic telangiectasia	2	protein kinase inhibitor activity	1	threonine endopeptidase activity	1	transferring hexosyl groups	1	skeletal development	1
energy derivation by oxidation of organic compounds	2	cyclin-dependent protein kinase inhibitor activity	1	peptidase activity	1	extracellular space	1	nervous system development	1
potassium channel activity	2	transforming growth factor beta receptor	1	identical protein-binding	1	glycogen metabolism	1	RASGRP1 (RAS guanyl releasing protein 1 (calcium and DAG-regulated))	
potassium ion transport	2	cytoplasmic mediator activity	1	anaphase-promoting complex-dependent proteasomal ubiquitin-dependent protein catabolism	1	protein C-terminus-binding	1	CLCN4 (chloride channel 4)	
negative regulation of cell proliferation	2	regulation of cyclin-dependent protein kinase activity	1	negative regulation of ubiquitin ligase activity during mitotic cell cycle	1	transcriptional repressor activity	1	MIA3 (melanoma inhibitory activity family member 3)	
carcinoma of colon	2	G1/S transition of mitotic cell cycle	1	positive regulation of ubiquitin ligase activity during mitotic cell cycle	1	DNA damage checkpoint	1	KCNK1 (potassium channel subfamily K member 1)	
protein transporter activity	2	induction of apoptosis	1	proteasome	1	G2 phase of mitotic cell cycle	1	CAMTA1 (calmodulin-binding transcription activator 1)	
protein import into nucleus	2	cell cycle arrest	1	proteasome degradation	1	negative regulation of transcription DNA-dependent	1	CDKN1B (cyclin-dependent kinase inhibitor 1B (p27 Kip1))	
ubiquitin-dependent protein catabolism	2	negative regulation of cell growth	1	proteasome pathway	1	response to endogenous stimulus	1	KPNB1 (karyopherin (importin) beta 1)	
hereditary diseases	2	negative regulation of phosphorylation	1	other connective tissue disease	1	transcription factor TFIIA complex	1	PSMA4 (proteasome (prosome macropain) subunit alpha 4)	
response to DNA damage stimulus	2	autophagic cell death	1	alzheimers disease	1	transcription coactivator activity	1	RCHY1 (ring finger and CHY zinc finger domain containing 1)	
unfolded protein-binding	2	ErbB signaling pathway	1	hepatitis b	1	transcription factor-binding	1	SMG1 (SMG1 homolog phosphatidylinositol 3-kinase-related kinase (C. elegans))	
acting on glycosyl bonds	2	prostate cancer	1	lupus erythematosus systemic	1	general RNA polymerase II transcription factor activity	1	EPM2AIP1 (EPM2A (laforin) interacting protein 1)	
usher syndrome	2	chronic myeloid leukemia	1	cardiomegaly	1	TATA-binding protein-binding	1	TOR1B (torsin family 1 member B (torsin B))	
RNA-binding	2	small cell lung cancer	1	cockayne syndrome	1	protein homodimerization activity	1	AGL (amylo-1 6-glucosidase 4-alpha-glucanotransferase)	
steroid hormone receptor activity	2	enzyme inhibitor activity	1	mixed connective tissue disease	1	protein heterodimerization activity	1	FOXN3_2 (forkhead box N3)	
nuclear receptors	2	regulation of protein kinase activity	1	retinoblastoma	1	transcription initiation from RNA polymerase II promoter	1	GTF2A2 (general transcription factor IIA 2 12kDa)	
protein complex assembly	2	receptor signaling protein activity	1	cell transformation neoplastic	1	RNA elongation from RNA polymerase II promoter	1	HNRNPA0 (heterogeneous nuclear ribonucleoprotein A0)	
via transesterification reactions with bulged adenosine as nucleophile	2	ovarian infertility genes	1	hypertrophy	1	positive regulation of transcription factor activity	1	NR1D2_1 (nuclear receptor subfamily 1 group D member 2)	
via transesterification reactions	2	regulation of transferase activity	1	werner syndrome	1	transcriptional preinitiation complex formation	1	NR1D2_2 (nuclear receptor subfamily 1 group D member 2)	
mRNA processing Reactome	2	regulation of cell proliferation	1	connective tissue diseases	1	positive regulation of specific transcription from RNA polymerase II promoter	1	SF3B3 (splicing factor 3b subunit 3 130kDa)	
membrane fraction	1	cell cycle-G1 to S control Reactome	1	ascites	1	general transcription from RNA polymerase II promoter	1	SFRS7 (splicing factor arginine/serine-rich 7 35kDa)	
endoplasmic reticulum membrane	1	IL-5	1	malignant neoplasms	1	basal transcription factors	1	ZNF91_2 (zinc finger protein 91)	
apparatus	1	PTEN dependent cell cycle arrest and apoptosis	1	ovarian cancer	1	RNA polymerase II transcription factor activity	1	GIPC3 (GIPC PDZ domain containing family member 3)	
plasma membrane	1	CDK regulation of DNA replication	1	mimicry	1	RNA transcription Reactome	1	GUSBP1 (glucuronidase beta pseudogene 1)	
guanyl-nucleotide exchange factor activity	1	cell cycle: G1/S check point	1	systemic scleroderma	1	autosomal recessive 3	1	KIAA0895 (KIAA0895)	
calcium ion-binding	1	cyclins and cell cycle regulation	1	virus diseases	1	autosomal recessive 18	1	R37702 (yf50d02.s1 Soares infant brain 1NIB Homo sapiens cDNA clone IMAGE:25627 3- mRNA sequence)	
diacylglycerol-binding	1	influence of ras and rho proteins on G1 to S transition	1	autoimmune diseases	1	retinal degeneration	1		
Ras protein signal transduction	1	regulation of p27 Phosphorylation during cell cycle progression	1	chronic liver disease	1	retinal diseases	1		
cell differentiation	1	CTCF: first multivalent nuclear factor	1	influenza	1	retinitis pigmentosa	1		

Figure 12 *CCRK*-repressive terms and occurrence numbers in high human left hemisphere based on GO, KEGG, GenMAPP, BioCarta and Disease via MAS 3.0. Ex, human left hemisphere; Inh, inhibition.

Discussion

CCRK-repressive different molecular Pearson mutual-positive-correlation network was setup in high human left hemisphere compared with the corresponding low chimpanzee left hemisphere (Fig. 11). Our identified *CCRK* inside-out inhibiting molecular network showed *CLCN4* (chloride channel 4), *MIA3* (melanoma inhibitory activity family member 3), *CDKN1B* (cyclin-dependent kinase inhibitor 1B (p27 Kip1)), *KPNB1* (karyopherin (importin) beta 1), *PSMA4* (proteasome (prosome macropain) subunit alpha 4), *SMG1* (*SMG1* homolog phosphatidylinositol 3-kinase-related kinase (C. elegans)), *AGL* (amylo-1 6-glucosidase 4-alpha-glucanotransferase) in high human left hemisphere. *CCRK* inside-out inhibiting terms network includes membrane, integral to membrane, nucleus, cytoplasm, cytosol, protein import into nucleus, NLS-bearing substrate import into nucleus, ribosomal protein import into nucleus, endomembrane system, nucleocytoplasmic transport, mechanism of protein import into the nucleus, mRNA export from nucleus, extracellular space; wound healing, autoimmune thyroid disease | thyroid disease, autoimmune, autoimmune diseases, response to DNA damage stimulus; role of ran in mitotic spindle regulation, positive regulation of leukocyte migration, negative regulation of cell adhesion, negative regulation of cell migration, G1/S transition of mitotic cell cycle, induction of apoptosis, cell cycle, cell cycle arrest, negative regulation of cell proliferation, negative regulation of cell growth, autophagic cell death, regulation of cell proliferation, cell cycle-G1 to S control Reactome, PTEN dependent cell cycle arrest and apoptosis, CDK regulation of DNA replication, cell cycle: G1/S check point, cyclins and cell cycle regulation, regulation of p27 Phosphorylation during cell cycle progression, negative regulation of ubiquitin ligase activity during mitotic cell cycle, positive regulation of ubiquitin ligase activity during mitotic cell cycle in high human left hemisphere based on integrative GO, KEGG, GenMAPP, BioCarta and disease databases (Fig. 12). Therefore, we propose high *CCRK* inside-out inhibiting DNA damage induced mitotic spindle through *CLCN4-MIA3-CDKN1B-KPNB1-PSMA4-SMG1-AGL* in human left hemisphere.

Spindle positive relationship with chloride channel, cyclin-dependent kinase inhibitor, importin has been reported in references as follows: 46 spindle configuration of in vitro-matured bovine oocytes exposed to sodium chloride or sucrose prior to cryotop vitrification; Triphenyltin chloride induces spindle microtubule depolymerisation and inhibits meiotic maturation in mouse oocytes [26, 27]. The prognosis in spindle-cell sarcoma depends on the expression of cyclin-dependent kinase inhibitor p27(Kip1) and cyclin E; Flavopiridol, a cyclin-dependent kinase inhibitor, prevents spindle inhibitor-induced endoreduplication in human cancer cells [28, 29]. Mitotic spindle scaling during Xenopus development by kif2a and importin alpha; Importin beta is transported to spindle poles during mitosis and regulates Ran-dependent spindle assembly factors in mammalian cells [30, 31].

In summary, *CCRK*-repressive molecular Pearson mutual-positive-correlation network was constructed in high human left hemisphere from the overlapping molecules of GRNInfer with Pearson. We propose and verify high *CCRK* inside-out inhibiting DNA damage induced mitotic spindle through *CLCN4-MIA3-CDKN1B-KPNB1-PSMA4-SMG1-AGL* in human left hemisphere. New critical insights and perspectives for the future will be further verified our hypothesis by experimental biology. High *CCRK* inside-out inhibiting DNA damage induced mitotic spindle through *CLCN4-MIA3-CDKN1B-KPNB1-PSMA4-SMG1-AGL* is very useful to identify novel markers and potential drugs for prognosis and therapy, develop a new route for studying the pathogenesis.

References

[1] Torikoshi, Y., K. Gohda, M.L. Davis, et al. Novel functional assay for spindle-assembly checkpoint by cyclin-dependent kinase activity to predict taxane chemosensitivity in breast tumor patient. J Cancer. [J] 4(9). 2013. 697-702.

[2] Simmons Kovacs, L.A., C.L. Nelson, and S.B. Haase. Intrinsic and cyclin-dependent kinase-dependent control of spindle pole body duplication in budding yeast. Mol Biol Cell. [J] 19(8). 2008. 3243-53.

[3] D'Angiolella, V., C. Mari, D. Nocera, et al. The spindle checkpoint requires cyclin-dependent kinase activity. Genes Dev. [J] 17(20). 2003. 2520-5.

[4] Neumayer, G., C. Belzil, O.J. Gruss, et al. TPX2: of spindle assembly, DNA damage response, and cancer. Cell Mol Life Sci. [J] 71(16). 2014. 3027-47.

[5] Wei, Y., H.T. Wang, Y. Zhai, et al. Mdb1, a fission yeast homolog of human MDC1, modulates DNA damage response and mitotic spindle function. PLoS One. [J] 9(5). 2014. e97028.

[6] Storey., J.D. A direct approach to false discovery rates. J. Roy. Stat. Soc., Ser. B. [J] 64. 2002. 479–498.

[7] Wang, Y., T. Joshi, X.S. Zhang, et al. Inferring gene regulatory networks from multiple microarray datasets. Bioinformatics. [J] 22(19). 2006. 2413-20.

[8] Wang, L., J. Huang, M. Jiang, et al. Activated PTHLH Coupling Feedback Phosphoinositide to

G-Protein Receptor Signal-Induced Cell Adhesion Network in Human Hepatocellular Carcinoma by Systems-Theoretic Analysis. ScientificWorldJournal. [J] 2012. 2012. 428979.

[9] Wang, L., J. Huang, M. Jiang, et al. Inhibited PTHLH downstream leukocyte adhesion-mediated protein amino acid N-linked glycosylation coupling Notch and JAK-STAT cascade to iron-sulfur cluster assembly-induced aging network in no-tumor hepatitis/cirrhotic tissues (HBV or HCV infection) by systems-theoretical analysis. Integr Biol (Camb). [J] 4(10). 2012. 1256-62.

[10] Wang, L., J. Huang, M. Jiang, et al. Tissue-specific transplantation antigen P35B (TSTA3) immune response-mediated metabolism coupling cell cycle to postreplication repair network in no-tumor hepatitis/cirrhotic tissues (HBV or HCV infection) by biocomputation. Immunol Res. [J] 52(3). 2012. 258-68.

[11] Wang, L., J. Huang, M. Jiang, et al. Signal transducer and activator of transcription 2 (STAT2) metabolism coupling postmitotic outgrowth to visual and sound perception network in human left cerebrum by biocomputation. J Mol Neurosci. [J] 47(3). 2012. 649-58.

[12] Lin, H., L. Wang, M. Jiang, et al. P-glycoprotein (ABCB1) inhibited network of mitochondrion transport along microtubule and BMP signal-induced cell shape in chimpanzee left cerebrum by systems-theoretical analysis. Cell Biochem Funct. [J] 30(7). 2012. 582-7.

[13] Huang, J., L. Wang, M. Jiang, et al. PTHLH coupling upstream negative regulation of fatty acid biosynthesis and Wnt receptor signal to downstream peptidase activity-induced apoptosis network in human hepatocellular carcinoma by systems-theoretical analysis. J Recept Signal Transduct Res. [J] 32(5). 2012. 250-6.

[14] Wang, L., L. Sun, J. Huang, et al. Cyclin-dependent kinase inhibitor 3 (CDKN3) novel cell cycle computational network between human non-malignancy associated hepatitis/cirrhosis and hepatocellular carcinoma (HCC) transformation. Cell Prolif. [J] 44(3). 2011. 291-9.

[15] Wang, L., J. Huang, M. Jiang, et al. AFP computational secreted network construction and analysis

between human hepatocellular carcinoma (HCC) and no-tumor hepatitis/cirrhotic liver tissues. Tumour Biol. [J] 31(5). 2011. 417-25.

[16] Wang, L., J. Huang, M. Jiang, et al. Survivin (BIRC5) cell cycle computational network in human no-tumor hepatitis/cirrhosis and hepatocellular carcinoma transformation. J Cell Biochem. [J] 112(5). 2011. 1286-94.

[17] Wang, L., J. Huang, M. Jiang, et al. MYBPC1 computational phosphoprotein network construction and analysis between frontal cortex of HIV encephalitis (HIVE) and HIVE-control patients. Cell Mol Neurobiol. [J] 31(2). 2011. 233-41.

[18] Wang, L., J. Huang, and M. Jiang. CREB5 computational regulation network construction and analysis between frontal cortex of HIV encephalitis (HIVE) and HIVE-control patients. Cell Biochem Biophys. [J] 60(3). 2011. 199-207.

[19] Wang, L., J. Huang, and M. Jiang. RRM2 computational phosphoprotein network construction and analysis between no-tumor hepatitis/cirrhotic liver tissues and human hepatocellular carcinoma (HCC). Cell Physiol Biochem. [J] 26(3). 2011. 303-10.

[20] Sun, L., L. Wang, M. Jiang, et al. Glycogen debranching enzyme 6 (*AGL*), enolase 1 (ENOSF1), ectonucleotide pyrophosphatase 2 (ENPP2_1), glutathione S-transferase 3 (*GSTM3*_3) and mannosidase (MAN2B2) metabolism computational network analysis between chimpanzee and human left cerebrum. Cell Biochem Biophys. [J] 61(3). 2011. 493-505.

[21] Huang, J.X., L. Wang, and M.H. Jiang. TNFRSF11B computational development network construction and analysis between frontal cortex of HIV encephalitis (HIVE) and HIVE-control patients. J Inflamm (Lond). [J] 7. 2011. 50.

[22] Sun, Y., L. Wang, M. Jiang, et al. Secreted Phosphoprotein 1 Upstream Invasive Network Construction and Analysis of Lung Adenocarcinoma Compared with Human Normal Adjacent Tissues by Integrative Biocomputation. Cell Biochem Biophys. [J] 56(2-3). 2010. 59-71.

[23] Huang, J., L. Wang, M. Jiang, et al. Interferon α-Inducible Protein 27 Computational Network Construction and Comparison between the Frontal Cortex of HIV Encephalitis (HIVE) and HIVE-Control Patients The Open Genomics Journal [J] 3(1875-693X). 2010. 1-8.

[24] Wang, L., Y. Sun, M. Jiang, et al. Integrative decomposition procedure and Kappa statistics for the distinguished single molecular network construction and analysis. J Biomed Biotechnol. [J] 2009. 2009. 726728.

[25] Wang, L., Y. Sun, M. Jiang, et al. FOS proliferating network construction in early colorectal cancer (CRC) based on integrative significant function cluster and inferring analysis. Cancer Invest. [J] 27(8). 2009. 816-24.

[26] Arcarons, N., R. Morato, J.F. Spicigo, et al. 46 spindle configuration of in vitro-matured bovine oocytes exposed to sodium chloride or sucrose prior to cryotop vitrification. Reprod Fertil Dev. [J] 27(1). 2014. 116.

[27] Shen, Y.T., Y.Q. Song, X.Q. He, et al. Triphenyltin chloride induces spindle microtubule depolymerisation and inhibits meiotic maturation in mouse oocytes. Reprod Fertil Dev. [J] 26(8). 2014. 1084-93.

[28] Goto, Y., S. Kawauchi, K. Ihara, et al. The prognosis in spindle-cell sarcoma depends on the expression of cyclin-dependent kinase inhibitor p27(Kip1) and cyclin E. Cancer Sci. [J] 94(5). 2003. 412-7.

[29] Motwani, M., X. Li, and G.K. Schwartz. Flavopiridol, a cyclin-dependent kinase inhibitor, prevents spindle inhibitor-induced endoreduplication in human cancer cells. Clin Cancer Res. [J] 6(3). 2000. 924-32.

[30] Wilbur, J.D. and R. Heald. Mitotic spindle scaling during Xenopus development by kif2a and importin alpha. Elife. [J] 2. 2013. e00290.

[31] Ciciarello, M., R. Mangiacasale, C. Thibier, et al. Importin beta is transported to spindle poles

during mitosis and regulates Ran-dependent spindle assembly factors in mammalian cells. J Cell Sci. [J] 117(Pt 26). 2004. 6511-22.

Chapter 4: High *BEST1*-inhibited nucleus actin cytoskeleton organization through *CAMTA1-EID1-PDE4DIP-NR1D2*

Abstract

40 different Pearson mutual-positive-correlation *BEST1*-repressive molecular network was constructed from 75 overlapping of 221 GRNInfer and 157 Pearson under *BEST1* CC ≤ -0.25 in high human left hemisphere compared with low chimpanzee left hemisphere. Our identified *BEST1*-inhibited nucleus molecular network showed *CAMTA1* (calmodulin-binding transcription activator 1), *EID1* (EP300 interacting inhibitor of differentiation 1), *PDE4DIP* (phosphodiesterase 4D interacting protein), *NR1D2* (nuclear receptor subfamily 1 group D member 2) in high human left hemisphere. *BEST1*-inhibited nucleus terms network includes cell cycle, cell differentiation, centrosome, actin cytoskeleton organization and biogenesis; transcription corepressor activity, protein-binding, specific transcriptional repressor activity, histone acetyltransferase regulator activity, histone acetyltransferase-binding, actin-binding, transcription factor activity, steroid hormone receptor activity, zinc ion-binding, sequence-specific DNA-binding, metal ion-binding, transcription, regulation of transcription DNA-dependent, cellular component, negative regulation of transcription from RNA polymerase II promoter, apparatus, actin filament-based process, actin cytoskeleton, circadian exercise, nuclear receptors based on integrative GO, KEGG, GenMAPP, BioCarta and disease databases in high human left hemisphere. Therefore, we propose high *BEST1*-inhibited nucleus actin

cytoskeleton organization through *CAMTA1-EID1-PDE4DIP-NR1D2* in human left hemisphere.

Keywords: *BEST1*-inhibited network; nucleus; actin cytoskeleton organization

Introduction

BEST1 is one of our identified significant molecules (fold change 2) in high human left hemisphere compared with the corresponding low chimpanzee left hemisphere. *BEST1* appears membrane fraction, cytosol, plasma membrane, integral to membrane, basolateral plasma membrane; ion channel activity, chloride channel activity, calcium ion-binding, chloride ion-binding; ion transport, visual perception, transepithelial chloride transport, response to stimulus; ion channel activity, visual perception, ion transport, sensory perception of light; ; visual loss, cystic fibrosis, drusen, vitelliform dystrophy, atrophy, age related macular degeneration, blindness, retinal diseases, macular dystrophy, vitelliform, vitreoretinochoroidopathy, maculopathy, Best Vitelliform Macular Dystrophy, macular degeneration, macular degeneration, age-related (AMD) based on GO, KEGG, GenMAPP, BioCarta and disease databases. Calcium-activated anion channels negative relationship with actin cytoskeleton organization has been reported in references as follows: The RGK family of GTP-binding proteins: regulators of voltage-dependent calcium channels and cytoskeleton remodeling; Disruption of actin cytoskeleton causes internalization of Ca(v)1.3 (alpha 1D) L-type calcium channels in salamander retinal neurons [1, 2]. Yet high *BEST1*-inhibited nucleus actin cytoskeleton organization through *CAMTA1-EID1-PDE4DIP-NR1D2* in human left hemisphere is not clear.

40 different Pearson mutual-positive-correlation *BEST1*-repressive molecular network was constructed from 75 overlapping of 221 GRNInfer and

157 Pearson under *BEST1* CC ≤ -0.25 in high human left hemisphere compared with low chimpanzee left hemisphere (Fig. 13A-13D).

A

Ex-Inh-GRNInfer
AA975427, ABCB1, ACADM, AF016004, AF052141, AL049278, AL049987, AL050030, ALDH7A1, ALG8, ANAPC10, ARF6, BCAS1, TUBGCP4_1, ABTB2, ADD3, ARHGAP12, ATP2B1, ATP5J, AW043812, BEST1, BMP2K, BRP44, BTN3A3, C14orf1, CA2, CACNB3, CALM1, CAMTA1, CCNO, CCRK, CD59, CDH19, COL6A1, CTBP1, CYFIP2, CYP2J2, DDX3Y, DHCR24, DLEU1, DTNA, DYNC1I1, DZIP3, EGFR, EID1, ENPP2_1, ENPP2_2, EPHX2, EXTL2, FAM114A1, FAM131B, FCMD, FEZ1, FKBPL, FLJ43806, FOXN3_1, GAPVD1, GBE1, GDF10, GEM, GM2A, GSTM3_2, GTF2A2, H24861, HERC2P3, HNRPH3, HSD17B6, HSPA9, IDI1, IFI44L, IMPA1, INHBB, KCNK1, KIAA0644, KIAA0888, L12535, LIMCH1, LOC157627, LOH11CR2A, LRPPRC, MAN2B2, MAP1B_1, MAP1B_2, MAP1B_3, MAPT, MED13, MEIS3P1, MFAP3L, MGC15523, MOAP1, MTMR15, NAIP, NFIB, NPC1, NR1D2_2, NUPR1, PALM2_AKAP2, PDE4DIP, PDE8A, PDIA2, PDLIM5, PGD, PHKG1, PPARD, RAB7L1, RBCK1, RCBTB2, RFK, RGL1, SAMHD1, SC4MOL, SEL1L, SELENBP1, SF3B3, SH3BGR, SLC13A3, SLC25A46, SMAD1_2, SPTLC1, SSH1, STAMBP, STAT2, STAU1, TH, THBD, TIGR:, TJP2, TMEM41B, TSPYL2, TULP3, U59632, UBR5, UBXD2, USP9Y, WASF3, WWOX, Z75311, ZNF174, ZNF254, ZNF294, C1orf61_1, CBLB, CDC25B, CDR2, CGRRF1, CHAD, CLDN10, CTNNA1_2, CTRL, DDX19A, DIXDC1, EVI5, FAM127A, FDFT1, FOXN3_2, GLOD4, GOLGA3, GSTM5, H10776, HCN2, HERC2P2, HNRNPA1, HNRPDL, HSP90AB1, HYPE, INSIG1, ITGB3BP, ITPR1, JARID1D, KIAA0368, KIF3A, L20971, MARCKSL1, MTMR1, MTUS1, NCK2, NDUFA5, NPAL3, NR1D2_1, OSBPL8, P4HB, PAK3, PCBP2_2, PHLDB1, PHYH, POLR2J, PON2, PRKRA, RAD50, RASGRP1, RNF14, RPL13, SCYE1, SFPQ_2, SFRS1, SGSH, SLC35E2, SMC5, SNRPE, SUB1, TAF1C, TFCP2, THBS2, TLOC1, TMEM63A, TRAPPC6A, TXNL4A, TYMS, U00928, UBB, UPF3A, USP22, USP46, USP8, UTY, W22289, W28620, W28807, WIPF2, ZNF271, ZNF443

B

Ex-Inh-Pearson
ELNR1, TUBGCP4_1, TUBGCP4_2, AA975427, AB016247, ACADM, ACTR2, AF052119, AGL, AK3P1, AL042668, AL050030, AL080234, AL109696, ANAPC10, ARF6, ATP2B1, ATP5J, ATP5J2, AW043812, B3GNT1, BAG5, BRP44, BTRC, C1D, CACNB3, CALM1, CAMTA1, CBLB, CDKN1B, CDR2, CGRRF1, CHAD, CLASP2, CLCN4, CLTB, COL6A1, CTBP1, CYFIP2, DCTN1, DDX19A, DGCR5, DICER1, DKFZp434H1419, DLEU1, DYNC1I1, DZIP3, EID1, EPM2AIP1, EXTL2, FEZ1, FOXN3_2, GAPVD1, GIPC3, GNAQ, GPR68, GTF2A2, GTF2I_1, H10776, HNRNPA0, HNRPDL, HOMER1, HSPA9, ISCA1, ITGB3BP, ITPR1, KCNK1, KIAA0368, KIAA0423, KIAA0888, KIAA0895, KIAA1109, KIF3A, KPNB1, LRPPRC, M19267, MAP1B_1, MAP1B_2, MAP1B_3, MAPT, MED13, MEIS3P1, MIA3, MOAP1, MTHFS, NDRG4, NDUFA5, NFIB, NPAL3, NPTN, NR1D2_1, NR1D2_2, OPTN, OSBPL8, PALM2_AKAP2, PCTK2, PDE4DIP, PER2, PIN1, POLR2J, POLR2J3, PPID, PPP1CA, PPP1R13B, PRDX2, PRKCI_1, PRKCI_2, PRKDC, PRKRA, PRKRIP1, PRPF19, PSMA4, R37702, RAB2A, RAB3GAP1, RAD50, RAD51C, RASGRP1, RBM34, RCHY1, RFK, RNF2, RPA3, RPP14, RPS26, SEL1L, SERINC3, SF3B3, SFRS1, SFRS7, SLC25A6, SLC35E2, SMAD1_2, SMG1, SPAG9, SPTB, SSH1, SSTR2, TERF1_1, TIPRL, TLOC1, TOR1B, TSPAN5, TULP3, U00928, U59632, U79289, USP46, USP8, VDAC2, W22289, W26407, WDFY3, WIPF2, ZNF443, ZNF91_1, ZNF91_2

C

Ex-Inh-Overlap-GRNInfer and Pearson
TUBGCP4_1, AA975427, ACADM, AL050030, ANAPC10, ARF6, ATP2B1, ATP5J, AW043812, BRP44, CACNB3, CALM1, CAMTA1, CBLB, CDR2, CGRRF1, CHAD, COL6A1, CTBP1, CYFIP2, DDX19A, DLEU1, DYNC1I1, DZIP3, EID1, EXTL2, FEZ1, FOXN3_2, GAPVD1, GTF2A2, H10776, HNRPDL, HSPA9, ITGB3BP, ITPR1, KCNK1, KIAA0368, KIAA0888, KIF3A, LRPPRC, MAP1B_1, MAP1B_2, MAP1B_3, MAPT, MED13, MEIS3P1, MOAP1, NDUFA5, NFIB, NPAL3, NR1D2_1, NR1D2_2, OSBPL8, PALM2_AKAP2, PDE4DIP, POLR2J, PRKRA, RAD50, RASGRP1, RFK, SEL1L, SF3B3, SFRS1, SLC35E2, SMAD1_2, SSH1, TLOC1, TULP3, U00928, U59632, USP46, USP8, W22289, WIPF2, ZNF443

D

Ex-Inh-Different Mutual Positive Pearson Correlation Compared with Con
ARF6, ATP2B1, CALM1, MAPT, RASGRP1, ATP5J, DDX19A, ITPR1, NDUFA5, NPAL3, KCNK1, SEL1L, TLOC1, CAMTA1, CDR2, DZIP3, EID1, HSPA9, MOAP1, OSBPL8, PDE4DIP, RFK, SFRS1, WIPF2, BRP44, MAP1B_1, MAP1B_2, CGRRF1, MEIS3P1, NR1D2_1, NR1D2_2, POLR2J, RAD50, USP8, KIF3A, TUBGCP4_1, AL050030, H10776, KIAA0888, W22289

Figure 13 **(A)** *BEST1*-repressive molecules of high human left hemisphere by GRNInfer. **(B)** *BEST1*-repressive molecules of high human left hemisphere by Pearson. **(C)** *BEST1*-repressive overlapping molecules of high human left hemisphere by GRNInfer and Pearson. **(D)** *BEST1*-repressive different mutual-positive-correlation molecules in high human left hemisphere compared with the corresponding low chimpanzee left hemisphere. Con, chimpanzee left hemisphere; Ex, human left hemisphere; Inh, inhibition.

Materials and Methods

441 significant molecules were identified from 12558 genes of 14 high human compared with 15 low chimpanzee left hemisphere in GEO data set GDS2678 (http://www.ncbi.nlm.nih.gov/sites/GDSbrowser?acc=GDS2678) containing brain cerebrum, anterior cingulated, anterior inferior parietal, anterior inferior temporal, middle frontal gyrus, frontal pole, etc. for studying high *BEST1*-inhibited nucleus actin cytoskeleton organization through *CAMTA1-EID1-PDE4DIP-NR1D2* using significant analysis of microarrays (SAM) (http://www-stat.stanford.edu/~tibs/SAM/) [3]. The raw microarray data were processed by log base 2. Two classes were unpaired and minimum fold change ⩾2 selected (the false-discovery rate 0%).

Gene expression values of *BEST1-repressive* different molecules were computed in high human left hemisphere compared with the corresponding low chimpanzee left hemisphere by AVERAGE and STDEV.

BEST1-repressive different mutual-positive-correlation molecular Pearson coefficients were computed in high human left hemisphere compared with the corresponding low no-tumor hepatitis/cirrhotic tissues under *BEST1* CC ≤ -0.25, as measurements of the correlation (linear dependence) including two variables X and Y and giving an inclusive value between −1 and +1.

BEST1-repressive molecular network was further constructed in high human left hemisphere by GRNInfer [4], GVedit tool and our articles [5-22]. GRNInfer is a novel mathematic method called GNR (Gene Network

Reconstruction tool) based on linear programming and a decomposition procedure for inferring gene networks. The method theoretically ensures the derivation of the most consistent network structure with respect to all of the datasets, thereby not only significantly alleviating the problem of data scarcity but also remarkably improving the reconstruction reliability. The following Equation (1) represents all of the possible networks for the same dataset. We established network based on the top 441 distinguished genes and selected parameters as lambda 0.0, threshold 1.0e-10.

$$J = (X' - A)U\Lambda^{-1}V^T + YV^T = \hat{J} + YV^T \qquad (1)$$

BEST1-repressive molecular knowledge network was further calculated in high human left hemisphere based on terms and occurrence numbers of GO (Cellular Component, Molecular Function and Biological Process), KEGG, GenMAPP, BioCarta and Disease via Molecule Annotation System, MAS (CapitalBio Corporation, Beijing, China; http://bioinfo.capitalbio.com/mas3/). The primary databases of MAS integrated various well-known biological resources, such as Gene Ontology (http://www.geneontology.org), KEGG (http://www.genome.jp/kegg/), BioCarta (http://www.biocarta.com/), GenMapp (http://www.genmapp.org/), HPRD (http://www.hprd.org/), etc.

Result

BEST1-repressive different Pearson mutual-positive-correlation molecular gene expression values were illustrated column diagrams by AVERAGE and STDEV in high human left hemisphere and the corresponding low chimpanzee left hemisphere, as shown in Fig. 14A.

BEST1-repressive different mutual-positive-correlation molecular Pearson coefficients were illustrated column diagrams in high human left hemisphere compared with the corresponding low chimpanzee left hemisphere, as shown in Fig. 14B and 14C, respectively.

BEST1-repressive molecular network was further constructed by GRNInfer in high human left hemisphere, as shown in Fig. 15.

BEST1-repressive knowledge terms network was further identified by MAS 3.0 in high human left hemisphere, as shown in Fig. 16.

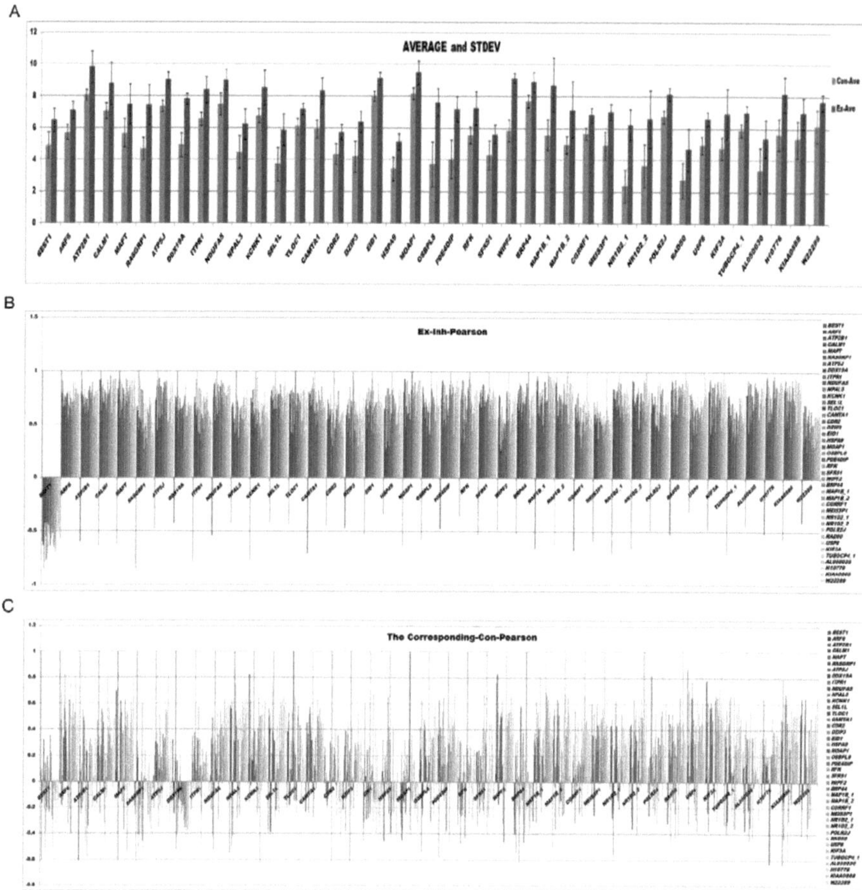

Figure 14 (A) Gene expression values of *BEST1*-repressive different mutual Pearson positive correlation molecules in high human left hemisphere and the corresponding low chimpanzee left hemisphere by AVERAGE and STDEV measurement. **(B)** Vertical quantification chart of *BEST1*-repressive different molecular mutual Pearson positive correlation coefficients in high human left hemisphere. n=14. **(C)** The corresponding correlation coefficients in low chimpanzee left hemisphere. n=15. Con, chimpanzee left hemisphere; Ex, human left hemisphere; Inh, inhibition.

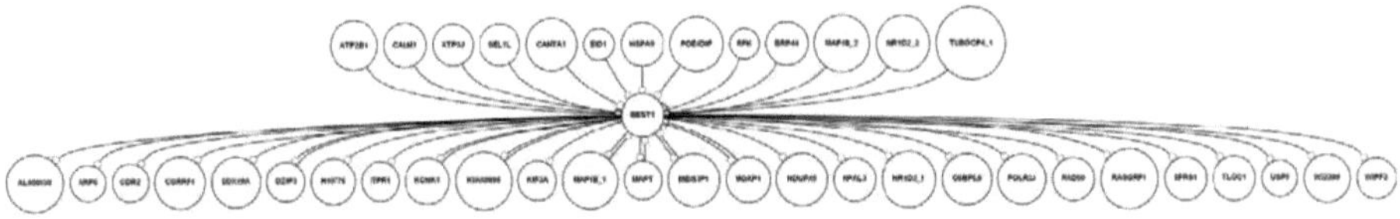

Figure 15 *BEST1*-repressive molecular network of high human left hemisphere by GRNInfer. n=14. Black arrow represents the activation relationship and empty circle as the inhibition one. Con, chimpanzee left hemisphere; Ex, human left hemisphere; Inh, inhibition.

Figure 16 *BEST1*-repressive terms and occurrence numbers in high human left hemisphere based on GO, KEGG, GenMAPP, BioCarta and Disease via MAS 3.0. Ex, human left hemisphere; Inh, inhibition.

Discussion

BEST1-repressive different molecular Pearson mutual-positive-correlation network was setup in high human left hemisphere compared with the corresponding low chimpanzee left hemisphere (Fig. 15).

Our identified *BEST1*-inhibited nucleus molecular network showed *CAMTA1* (calmodulin-binding transcription activator 1), *EID1* (EP300 interacting inhibitor of differentiation 1), *PDE4DIP* (phosphodiesterase 4D interacting protein), *NR1D2* (nuclear receptor subfamily 1 group D member 2) in high human left hemisphere. *BEST1*-inhibited nucleus terms network includes cell cycle, cell differentiation, centrosome, actin cytoskeleton organization and biogenesis; transcription corepressor activity, protein-binding, specific transcriptional repressor activity, histone acetyltransferase regulator activity, histone acetyltransferase-binding, actin-binding, transcription factor activity, steroid hormone receptor activity, zinc ion-binding, sequence-specific DNA-binding, metal ion-binding, transcription, regulation of transcription DNA-dependent, cellular component, negative regulation of transcription from RNA polymerase II promoter, apparatus, actin filament-based process, actin cytoskeleton, circadian exercise, nuclear receptors in high human left hemisphere based on Integrative GO, KEGG, GenMAPP, BioCarta and disease databases (Fig. 16). Therefore, we propose high *BEST1*-inhibited nucleus actin cytoskeleton organization through *CAMTA1-EID1-PDE4DIP-NR1D2* in human left hemisphere.

Actin cytoskeleton organization positive relationship with calmodulin has been reported in reference as follows: Calmodulin controls organization of the actin cytoskeleton via regulation of phosphatidylinositol (4,5)-bisphosphate synthesis in Saccharomyces cerevisiae [23].

In summary, *BEST1*-repressive molecular Pearson mutual-positive-correlation network was constructed in high human left hemisphere from the overlapping molecules of GRNInfer with Pearson. We propose and verify high *BEST1*-inhibited nucleus actin cytoskeleton organization through *CAMTA1-EID1-PDE4DIP-NR1D2* in human left hemisphere. New critical insights and perspectives for the future will be further verified our hypothesis by experimental biology. High *BEST1*-inhibited nucleus actin cytoskeleton organization through *CAMTA1-EID1-PDE4DIP-NR1D2* is very useful to identify novel markers and potential drugs for prognosis and therapy, develop a new route for studying the pathogenesis.

References

[1] Correll, R.N., C. Pang, D.M. Niedowicz, et al. The RGK family of GTP-binding proteins: regulators of voltage-dependent calcium channels and cytoskeleton remodeling. Cell Signal. [J] 20(2). 2008. 292-300.

[2] Cristofanilli, M., F. Mizuno, and A. Akopian. Disruption of actin cytoskeleton causes internalization of Ca(v)1.3 (alpha 1D) L-type calcium channels in salamander retinal neurons. Mol Vis. [J] 13. 2007. 1496-507.

[3] Storey., J.D. A direct approach to false discovery rates. J. Roy. Stat. Soc., Ser. B. [J] 64. 2002. 479–498.

[4] Wang, Y., T. Joshi, X.S. Zhang, et al. Inferring gene regulatory networks from multiple microarray datasets. Bioinformatics. [J] 22(19). 2006. 2413-20.

[5] Wang, L., J. Huang, M. Jiang, et al. Activated PTHLH Coupling Feedback Phosphoinositide to G-Protein Receptor Signal-Induced Cell Adhesion Network in Human Hepatocellular Carcinoma by Systems-Theoretic Analysis. ScientificWorldJournal. [J] 2012. 2012. 428979.

[6] Wang, L., J. Huang, M. Jiang, et al. Inhibited PTHLH downstream leukocyte adhesion-mediated

protein amino acid N-linked glycosylation coupling Notch and JAK-STAT cascade to iron-sulfur cluster assembly-induced aging network in no-tumor hepatitis/cirrhotic tissues (HBV or HCV infection) by systems-theoretical analysis. Integr Biol (Camb). [J] 4(10). 2012. 1256-62.

[7] Wang, L., J. Huang, M. Jiang, et al. Tissue-specific transplantation antigen P35B (TSTA3) immune response-mediated metabolism coupling cell cycle to postreplication repair network in no-tumor hepatitis/cirrhotic tissues (HBV or HCV infection) by biocomputation. Immunol Res. [J] 52(3). 2012. 258-68.

[8] Wang, L., J. Huang, M. Jiang, et al. Signal transducer and activator of transcription 2 (STAT2) metabolism coupling postmitotic outgrowth to visual and sound perception network in human left cerebrum by biocomputation. J Mol Neurosci. [J] 47(3). 2012. 649-58.

[9] Lin, H., L. Wang, M. Jiang, et al. P-glycoprotein (ABCB1) inhibited network of mitochondrion transport along microtubule and BMP signal-induced cell shape in chimpanzee left cerebrum by systems-theoretical analysis. Cell Biochem Funct. [J] 30(7). 2012. 582-7.

[10] Huang, J., L. Wang, M. Jiang, et al. PTHLH coupling upstream negative regulation of fatty acid biosynthesis and Wnt receptor signal to downstream peptidase activity-induced apoptosis network in human hepatocellular carcinoma by systems-theoretical analysis. J Recept Signal Transduct Res. [J] 32(5). 2012. 250-6.

[11] Wang, L., L. Sun, J. Huang, et al. Cyclin-dependent kinase inhibitor 3 (CDKN3) novel cell cycle computational network between human non-malignancy associated hepatitis/cirrhosis and hepatocellular carcinoma (HCC) transformation. Cell Prolif. [J] 44(3). 2011. 291-9.

[12] Wang, L., J. Huang, M. Jiang, et al. AFP computational secreted network construction and analysis between human hepatocellular carcinoma (HCC) and no-tumor hepatitis/cirrhotic liver tissues. Tumour Biol. [J] 31(5). 2011. 417-25.

[13] Wang, L., J. Huang, M. Jiang, et al. Survivin (BIRC5) cell cycle computational network in human

no-tumor hepatitis/cirrhosis and hepatocellular carcinoma transformation. J Cell Biochem. [J] 112(5). 2011. 1286-94.

[14] Wang, L., J. Huang, M. Jiang, et al. MYBPC1 computational phosphoprotein network construction and analysis between frontal cortex of HIV encephalitis (HIVE) and HIVE-control patients. Cell Mol Neurobiol. [J] 31(2). 2011. 233-41.

[15] Wang, L., J. Huang, and M. Jiang. CREB5 computational regulation network construction and analysis between frontal cortex of HIV encephalitis (HIVE) and HIVE-control patients. Cell Biochem Biophys. [J] 60(3). 2011. 199-207.

[16] Wang, L., J. Huang, and M. Jiang. RRM2 computational phosphoprotein network construction and analysis between no-tumor hepatitis/cirrhotic liver tissues and human hepatocellular carcinoma (HCC). Cell Physiol Biochem. [J] 26(3). 2011. 303-10.

[17] Sun, L., L. Wang, M. Jiang, et al. Glycogen debranching enzyme 6 (*AGL*), enolase 1 (ENOSF1), ectonucleotide pyrophosphatase 2 (ENPP2_1), glutathione S-transferase 3 (*GSTM3*_3) and mannosidase (MAN2B2) metabolism computational network analysis between chimpanzee and human left cerebrum. Cell Biochem Biophys. [J] 61(3). 2011. 493-505.

[18] Huang, J.X., L. Wang, and M.H. Jiang. TNFRSF11B computational development network construction and analysis between frontal cortex of HIV encephalitis (HIVE) and HIVE-control patients. J Inflamm (Lond). [J] 7. 2011. 50.

[19] Sun, Y., L. Wang, M. Jiang, et al. Secreted Phosphoprotein 1 Upstream Invasive Network Construction and Analysis of Lung Adenocarcinoma Compared with Human Normal Adjacent Tissues by Integrative Biocomputation. Cell Biochem Biophys. [J] 56(2-3). 2010. 59-71.

[20] Huang, J., L. Wang, M. Jiang, et al. Interferon α-Inducible Protein 27 Computational Network Construction and Comparison between the Frontal Cortex of HIV Encephalitis (HIVE) and HIVE-Control Patients The Open Genomics Journal [J] 3(1875-693X). 2010. 1-8.

[21] Wang, L., Y. Sun, M. Jiang, et al. Integrative decomposition procedure and Kappa statistics for the distinguished single molecular network construction and analysis. J Biomed Biotechnol. [J] 2009. 2009. 726728.

[22] Wang, L., Y. Sun, M. Jiang, et al. FOS proliferating network construction in early colorectal cancer (CRC) based on integrative significant function cluster and inferring analysis. Cancer Invest. [J] 27(8). 2009. 816-24.

[23] Desrivieres, S., F.T. Cooke, H. Morales-Johansson, et al. Calmodulin controls organization of the actin cytoskeleton via regulation of phosphatidylinositol (4,5)-bisphosphate synthesis in Saccharomyces cerevisiae. Biochem J. [J] 366(Pt 3). 2002. 945-51

Acknowledgments

This work was supported by the National Natural Science Fund (61171114) and Key Fund (61433015), the National Social Science Major Fund (14ZDB154) of China, Fundamental Research Funds for the Central Universities (BUPT Project No: 2014RC0201), Research Innovation Fund of Beijing University of Posts and Telecommunications (BUPT Project No: 2014XD-01), Research Innovation Fund for College Students of Beijing University of Posts and Telecommunications.

FSC
www.fsc.org
MIX
Papier aus verantwortungsvollen Quellen
Paper from responsible sources
FSC® C105338

Printed by Books on Demand GmbH, Norderstedt / Germany